THE
Berkshire
GLASS WORKS

THE
Berkshire
GLASS WORKS

WILLIAM J. PATRIQUIN & JULIE L. SLOAN

Charleston | London

THE
History
PRESS

Published by The History Press
Charleston, SC 29403
www.historypress.net

Copyright © 2011 by William J. Patriquin and Julie L. Sloan
All rights reserved

First published 2011

Manufactured in the United States

ISBN 978.1.60949.282.3

Library of Congress CIP data applied for.

CONTENTS

ACKNOWLEDGEMENTS

No research project is possible without the help of many other people. We would like to thank especially Annemarie Leone and the staff of the Berkshire Atheneum Reference Department. They never failed to find a reference or source whenever we asked. Their patience and knowledge were invaluable. Others who provided access to their homes, collections and knowledge include Gayle Bardhan, Rakow Library, Corning Museum of Glass; David Bondini; Mary Carrow; Anne Cathcart and Donna Hasler, Chesterwood, Lenox, Massachusetts; Angie Chase; Dean and Kathy Clement; Jerre Croteau, Grace Episcopal Church, New Bedford, Massachusetts; Lori Dilego; Charlie Flint, Charles Flint Antiques, Lenox, Massachusetts; Katherine A. Gardner-Westcott, Watertown Free Public Library, Watertown, Massachusetts; Linda Hall, Williams College Archives, Williamstown, Massachusetts; Leann Hayden, Berkshire Museum, Pittsfield, Massachusetts; Peter Houghton, Waterloo Library and Historical Society, Waterloo, New York; Ed Kirby; Ed and Kim LaMarre; Lawrence and Nancy LaMarre; the late Jane Manning; Jeanne Maschino, librarian, *Berkshire Eagle*, Pittsfield, Massachusetts; Gerry Robichaud; Roberto Rosa, Serpentino Stained and Leaded Glass, Needham Heights, Massachusetts; Karen Rosado and Lyn Hovey, Lyn Hovey Studio, Hyde Park, Massachusetts; Betsy Sherman and the staff of the Berkshire Historical Society, Pittsfield, Massachusetts; Leo and Ann Sondrini; James Yarnall; Kevin and Sandy White; Peggy Williams; and Dave and Sherri Wilson. We apologize to anyone we may have omitted—in a project this long, we are bound to have a few "senior moments."

Acknowledgements

Special thanks to Bill's wife, Kathy Patriquin, for tolerating his obsession with glass for all these years, and to the property owners of Berkshire Village who let him dig up their property—you know who you are. Special thanks, too, to Julie's husband, Albert Lewis, for everything.

INTRODUCTION

A cigar box filled with shards of colored glass sits on the table. As the sunlight hits it, glints of burgundy, cobalt, teal, opaline and deep, mossy green play around the box and on the tabletop. Dug from middens in friends' backyards, most pieces were found a few dozen feet from the ruins of a glass furnace. This and a few other artifacts are all that remain of a once-thriving business, the Berkshire Glass Works, the first glass factory in the United States to make colored cathedral glass and one of the earliest to blow antique stained glass.

Berkshire County, often called simply "The Berkshires," is at the westernmost end of the state of Massachusetts, stretching from Connecticut to Vermont, abutting New York State. The Hoosac Mountains, part of the ancient Appalachian range, run from north to south in gently undulating hills. The county seat, Pittsfield, lies at the geographical center of the county, dividing "North County" from "South County." The town of Lanesborough is immediately north of Pittsfield, and Cheshire is to the northeast of Lanesborough. The hamlet variously called East Lanesborough, Berkshire and Berkshire Village sits in the southeast corner of Lanesborough on the border of Cheshire. In the nineteenth century, the Berkshires looked to the west, to the Hudson River, for its physical connection to the outside world. In 1841, the Boston & Albany Railroad built a line from Pittsfield west to the port at Hudson, New York (forty-five miles away), called the Hudson and Berkshire Railroad. Another rail line to North Adams (about twenty-two miles north of Pittsfield) connected the northern towns and cities to Pittsfield in 1846, running through East Laneborough at Berkshire Village.[1]

View of Berkshire Village from the north. The Hoosac Mountain range is on the left, and the Taconic range is just visible on the right. *Courtesy of the Berkshire Atheneum, Pittsfield, Massachusetts.*

Glass is made primarily of silica, found in nature as sand, with alkalis—soda, lime and potash—added to lower the melting point of the sand and make it easier to manipulate. Northern Berkshire County has long been known for the extraordinary purity of its sand, which made it essential to glass manufacturers around the country in the nineteenth century. With the source of the main ingredient so close at hand, a number of enterprising businessmen tried their hands at glass, making the county an important glass center until the turn of the twentieth century.

One of those manufacturers was located in Lanesborough. Known as the Berkshire Glass Works, it survived for slightly more than fifty years. At its peak, the factory employed over two hundred people and produced window glass, plate glass and colored glass for stained-glass windows. The production of clear glass was not unique or even unusual—just down the road from Berkshire Village, Lenox Plate Glass Works did the same thing, while an early factory in Cheshire produced crown glass in the first decade of the nineteenth century. But the production of colored glass *was* unusual, at least in the 1870s.

Part I

THE HISTORY
OF THE BERKSHIRE
GLASS WORKS

SAND

Quartz sand (silica) is the primary ingredient of glass. The sands of Berkshire County are 97 to 99.6 percent pure silica. Such purity is rare. Berkshire County sand was formed during the Cambrian Age and later. Beaches composed of sand, crushed rock, decomposing plants and marine life were deposited along the shores of the Iapetus Ocean—roughly the same area as today's Atlantic Ocean—some 570 million years ago.[2] These deposits were subsequently buried under several thousand feet of sediment during the next several million years. As the continents relocated, the heat resulting from tectonic plate movement burned off the organic components of the buried beaches and metamorphosed the sand into quartzite, a stone of pure silica. During the last ice age, which ended about 10,000 years ago, the proglacial Lake Bascom formed between the present-day Taconic range in eastern New York and the Hoosac range in western Massachusetts, extending into southern Vermont. Slightly south of Pittsfield, in the town of Washington, the quartz is layered with mica, slate and clay, making it very hard and capable of withstanding high heat, so that it was suitable for the hearths and walls of blast furnaces for smelting iron ore, another major industry of northern Berkshire County.[3] North of Pittsfield, the movement of the thick ice and melt water ground the quartzite into sand that was fine and pure, with grains that are angular and sharp—the best for glassmaking— depositing it at the lower reaches of the western slopes of the Hoosac Range.[4] Beds of sand are found along the eastern side of present-day State Route 8 in Lanesborough and Cheshire. In locations where melt water moved slowly, only silt and small quartz grains were carried and deposited. Other

Left: Crystal Hill, Berkshire Village, quartzite.
Right: Lane bed sand. *Julie L. Sloan.*

beds contain a variety of particle sizes ranging from silt and small particles to pebbles, stones, larger rocks and boulders.[5] The veins of sand are vertical and close to the surface, easily mined with picks.[6]

Sand for glassmaking has been mined in this region since around 1806.[7] Cheshire was recognized early on as having very fine glass sand located in what was called the Lane bed. A geologist noted in 1832 that the quartz in Cheshire "is so much disintegrated, that it easily crumbles into a beautiful white sand," perfect for glassmaking. It was sold for six and a quarter cents per bushel and used in glass factories in Cheshire; Warwick, Massachusetts; Utica, New York; and Keene, New Hampshire.[8] Cheshire sand was sent to factories as far away as Boston in 1812 or earlier.

Sand from the Lane bed was good, but in 1845 better sand was discovered in Cheshire beneath the ruins of the 1812 Cheshire Crown Glass factory, ironically. This sand was exceptional, and its discovery quickly led to the founding of the Berkshire Glass Company. The vein was said to have been discovered when the North Adams rail line was laid. An adjacent bed was discovered when excavating for a gristmill and another when digging wells.[9] The sand here was only ten feet below grade, and the clay and dirt around it were not hard, so excavation was easy.[10] The quarry was "of dazzling whiteness, in the midst of the dark evergreens by its side."[11] The property owner, Daniel B. Brown III, sent samples of the sand to Boston to be analyzed and was informed of its extraordinarily high purity—better than that from the Lane bed—that could create a glass that was almost colorless.[12] (Most clear glass has a slight green tint, easily visible at the edges, caused by iron in the sand.) By 1856, it was written that "nearly all manufactories of nice glass in the world are now supplied" with Berkshire sand, including the famous Boston & Sandwich factory and as far away as Liverpool, England, and Le Havre, France.[13] The renowned English glassworks of Thomas Webb & Sons, Stourbridge, called Berkshire sand "the *finest* we have *ever* used."[14] The history of the sand industry in this area is beyond the scope of this work but is nevertheless interesting.[15]

BERKSHIRE GLASS COMPANY, 1847–1858

Samuel Smith (dates unknown), who resided in Boston but was a native of Cheshire and a landowner there, learned of the discovery of veins of high-quality glass-making sand in the vicinity of Lanesborough and Cheshire from Daniel Brown III in 1845. He began buying up property there and in neighboring towns.[16] Between February and May 1847, Smith acquired title to most of the land where there was sand in Cheshire and in the part of the Lanesborough that would become Berkshire Village.[17] William D.B. Linn (circa 1807–1870), a dealer of marble in Pittsfield, also bought some properties in the area to use the sand to cut marble. Smith and Linn, along with Lanesborough lawyer and justice of the peace William T. Filley (b. 1817), incorporated the Berkshire Glass Company in 1847.[18]

Initially, the company was a means to control the sand beds of Cheshire and Lanesborough—a factory was not constructed until 1853. The principals (who changed almost every year) and investors were all businessmen in Boston and New York, although some had local connections. Stock was actively traded between 1847 and 1853.[19]

It is not clear what spurred the company to finally erect a glass factory, housing and subsidiary businesses in 1853. Lack of construction at an earlier date may have been related to funding—perhaps Smith, Linn and Filley had spent their available capital on land acquisitions and needed to raise more through the sale of stock before they could build. Perhaps more important was the fact that none of these men were glassmakers. In late 1852, a fortuitous tragedy brought them the technical expertise they lacked. A glass factory in Sand Lake, New York, about twenty-five miles

View of a glass factory. *From* Harper's Weekly, *January 12, 1884.*

west of Lanesborough, burned to the ground on Christmas that year. In business since about 1806 as the Rensselaer Glass Factory, it was owned by brothers Albert R. (1810–1896) and Samuel H. Fox (dates unknown), who had purchased it from their father, Isaac, and a partner, Nathan Crandall, in 1838. The fire so devastated the factory that it was never reopened.[20] The Berkshire Glass Company snapped up Albert Fox to become its manager. Fox brought equipment as well as knowledge and workers from Sand Lake to Lanesborough.[21] Under Fox's administration, the building of the glassworks and its supporting village was completed in eleven months, "as if by magic." A post office opened, naming the hamlet "Berkshire." The newspaper predicted that "East Lanesborough [Berkshire] is destined largely to increase in population and wealth."[22]

The initial product of the company was common window glass, blown in the cylinder method. To make glass, a man called a *teazer* mixed the powdered silica, alkalis and other elements together to form a *batch*. The batch was placed into the clay *pot* inside a *furnace* to melt, a process that typically took thirteen to sixteen hours. Molten glass was referred to as *metal*. At Berkshire the rectangular furnace was situated in the center of the glass house. Its platform, called the *bench*, on which the men worked was about eight feet from the ground, leaving room below the pots for the fire. In the cylinder

The History of the Berkshire Glass Works

Above: A gatherer. He has just taken the hot lump of glass from the pot on the end of a blowpipe. He is pouring water over the stem of the blowpipe to cool it down. Around his neck hangs a wooden mask to protect his face from the heat of the furnace when he reaches into it for the gather. *From* Harper's New Monthly Magazine, *"A Piece of Glass," July 1889.*

Right: Gatherer's mask from Berkshire Glass Works with mouthpiece. The gatherer held the mask in place by gripping the mouthpiece with his teeth. *Courtesy of Charles Flint.*

Left: The initial inflation of the gather. *From* Harper's New Monthly Magazine, *"A Piece of Glass," July 1889.*

Below: The blower puts the growing cylinder back into the furnace at regular intervals to keep it hot enough to manipulate. *From* Harper's New Monthly Magazine, *"A Piece of Glass," July 1889.*

Right: Capping the cylinder. In the background at the center are blowpipes. On the right is a gatherer wearing his mask. *From* Harper's New Monthly Magazine, *"A Piece of Glass," July 1889.*

Below: Splitting the cylinders and placing them into the flattening oven. *From* Harper's Weekly, *January 12, 1884.*

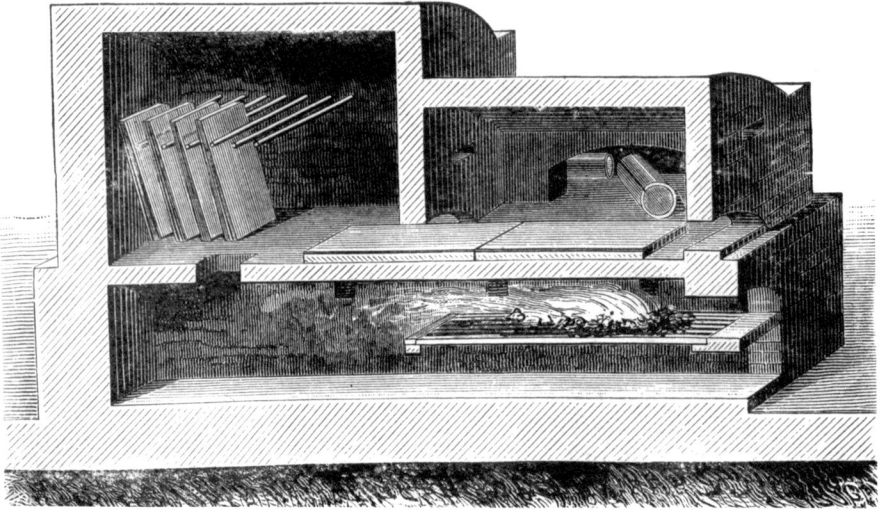

Flattening and annealing oven. The cylinders are reheated on the right and moved toward a flat bed, in the foreground, where they are opened and flattened by a man reaching into the oven with a long stick with a board on the end. On the left side, the flattened sheets of glass stand in a rack as they cool. *From Charles Annandale, ed.*, The Popular Encyclopedia or, Conversations Lexicon, *vol. 6, pl LXXXVII-LXXXIX (London: Blackie & Son, 1883).*

method of making window glass, the molten glass was gathered onto a blow pipe by a worker called a *gatherer*. The gatherers wore masks to protect their faces from the heat of the furnace. The mask was made of a flat plank of wood with a tab to take between their teeth to hold the mask in place. The holes for the eyes were covered with blue glass to filter the harmful rays of light emanating from the molten glass. Three or four gathers were made to get enough molten glass, weighing between eleven and twenty-five pounds, on the end of the pipe to make one sheet.[23] After making the first inflation of the bubble, the gatherer passed the pipe to the *blower*. The blower alternately blew into the pipe to inflate the bubble and swung it like a pendulum below the level of the bench, sometimes even completing the arc above his head to force the bubble to become elongated. The blower frequently had to reheat the bubble in the furnace to keep the metal malleable.

When it reached the appropriate length (about six feet), the bubble was opened at the bottom and then cracked off the pipe and allowed to cool enough to handle with insulated mitts. The cylinder (also called a *muff* or a *roller*) was placed horizontally on a wooden horse and marked by the blower, who was paid by the number of cylinders he produced. At this point, the cylinders were shaped like a bottomless bottle, with the curved top where it was closest to the blowpipe still attached. It was the *capper's* job to cut off these rounded shoulders in a process called *capping*. Following a mark made by the blower, the capper took a small knob of molten glass and wrapped it around the cylinder like a string. A sharp rap with a cold iron rod caused the glass to break at this point in a neat seam. The curved top of the cylinder fell into a waste bin to be remelted. Such waste glass was called *cullet*; the term also included the leftover glass in the melting pots and on the blowpipes. The capper then split the cylinder lengthwise with a diamond cutter or a hot iron rod. The cut cylinder was then moved to the *flattening oven*, where the glass was reheated to become flexible. The *flattener* reached into the oven through a small hole and coaxed the cylinder open with a rod. He then flattened and smoothed it into a sheet using a wooden bar attached to a long pipe. The glass then moved into an *annealing oven* or *leer* (or *lehr*), where it was gradually cooled to room temperature. If it was cooled too quickly, it could break or become unstable and break later. The cooled sheets were then cut into regularly sized panes and packed to ship.[24]

The first cylinders were blown at Berkshire Glass Company in November 1853.[25] By the following month, the glassworks was described as "unsurpassed by any…at home or abroad."[26] Within the next month (January 1854), the firm sold $8,000 worth of glass ($212,000 today). By 1855, the glass factory was making three thousand feet per day.[27] In the fall of that year, its cylinder glass won a local prize for its superior quality.[28] It was sold sometimes as Berkshire crystal. The company also produced blown-glass bell jars for clocks and other display items.[29]

It is not clear exactly how long Albert Fox continued as the manager, but it was a short time. William G. Harding, a later owner, wrote that Fox stayed only one year, although he is described as the manager in an 1855 publication.[30] He is not mentioned after that point. Stephen T. Whipple took over the role of manager until about 1857.[31] Whipple was a local landowner who would play a sporadic role throughout the history of the Berkshire Glass Company, buying and selling property to and for it.

From the annual financial notices published between 1854 and 1857, it would appear that the factory was a monetary success, but it, like thousands of other American businesses, fell prey to the financial panic of 1857, which occurred in August as a result of a confluence of international events.[32] The factory closed.[33] The following month, the glass company published its annual statement of value on the same days that it ran advertisements for the sale of the factory as "extensive" and "valuable."[34] The glass factory had one eight-pot furnace with structures that were "spacious [and] well arranged." Included in the sale were a boardinghouse; a few small houses for employees; all machinery; the twelve acres of land on which the factory sat; and an additional five hundred acres of woodland that supplied the fuel for the furnaces. The advertisement optimistically noted that there was room for an additional glass house. It stressed that access was available to Albany (fifty-five miles away) and Hudson, New York, from the Pittsfield & North Adams Railroad. The "privilege of digging sand...the quality of which is unequalled" was also part of the deal. George W. Gordon (d. 1877), treasurer of the glass company and future owner of the Cheshire Sand Works, was accepting offers at his office in Boston.

Six months later, in March 1858, the *Pittsfield Sun* reported that the company had been sold to James N. Richmond, a local businessman who had previously been associated with the plate glassworks in neighboring Cheshire and Lenox. Richmond purchased the works for $82,000 ($2.2 million today).[35] Then, barely four months after he had purchased it, he sold it for only $20,000 ($540,000 today) to Page & Robbins, a Boston glass dealer, in July. The reasons for selling at such a loss are unknown but were hinted at by Thomas Gaffield (1825–1900), a glassmaker from Boston who kept a detailed journal about the glass industry. Gaffield wrote that Richmond and Stephen Whipple were "managing, or mis-managing" the factory.[36]

Page & Robbins was a dealer of French and German window glass, Lenox plate glass and glazing supplies. Harrison P. Page and James Robbins of Boston were the owners.[37] Finally, the glassworks had found an owner willing to put time, effort and confidence into the company. Although its history would be unsteady, like that of most glass houses, Harrison Page (1815–1898) would own the company until his death in 1898.

Page & Robbins, 1858–1863

Harrison P. Page

Thomas Gaffield described Harrison Page in admiring terms:

> *No man in our country has labored with greater zeal and more unremitting industry to advance the perfection of the sheet glass manufacture in our country than Mr. [Harrison] Page. In season and out of season, he has been found at the Berkshire Glass Works superintending the digging of the pure white sand which enters into its composition, watching the progress of the pot-maker's work, the mixture of the batch, the making of the teazer, the founder, the gatherer, and the blower in the furnace-room, and all the nice manipulations and careful movements which are so absolutely necessary to ensure the production of a sheet of glass free from impurities of color and imperfection of body and surface. He deserves the praise of all who are interested practically or scientifically, in the advance of American industry and the desire to make our country free from the necessity of importing any glass from abroad.*[38]

Born in Augusta, Maine, little is known about Page's background. By 1856, he had become the partner of James Robbins in a store selling imported window glass, located at 11 Broad Street or 5 State Street Block in Boston.[39] Robbins was a member of the Boston Board of Trade in that year.[40] Page did not become a member until 1866, when Robbins left the firm.[41]

Page lived in Watertown, Massachusetts, a small town outside of Boston. Although he signed petitions for local land-use issues and occasionally gave

Harrison P. Page's house, Galen Street, Watertown, Massachusetts, demolished in 1895. *Courtesy of the Watertown, Massachusetts Free Public Library.*

donations, he was locally prominent because of his wealth, not because of his activities.[42] Despite this low-key appearance, however, Page was repeatedly described as "energetic" when it came to glassmaking and his factory in Berkshire, which he visited often.[43]

* * *

One of Page & Robbins' first acts, in October 1858, was to sell the sand beds of the Berkshire Glass Company to George W. Gordon, the former treasurer of the glass company, for $10,000 ($270,000 today), immediately recouping half of their initial investment.[44] They instantly sank some $20,000 into the plant, building new homes for workers and improving the factory.[45] Having retained rights to the Cole sand bed in Lanesborough, the factory also dug its own raw material. For this one brief period of time, until about 1863 when it began to purchase coal, the factory had both sand and fuel (the surrounding forests) in ready and close supply. Its window glass received accolades for its "excellence" and "perfection," winning a silver medal at the Ninth Exhibition of the Massachusetts Charitable Mechanic Association in Boston in 1860.[46]

Page & Robbins hired James Raybold (1810–1869), an Englishman, to oversee the works. Raybold's son (also James) was a pot maker. When the

Right: An 1858 map of East Lanesborough. *Map of the County of Berkshire, Massachusetts, Smith Gallup & Co., 1858. Courtesy of the Berkshire Atheneum, Pittsfield, Massachusetts.*

Below: View of Berkshire Village, circa 1860. *Courtesy of the Berkshire Atheneum, Pittsfield, Massachusetts.*

25

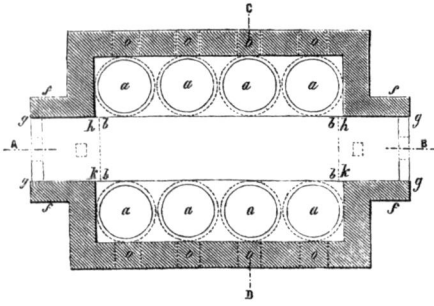

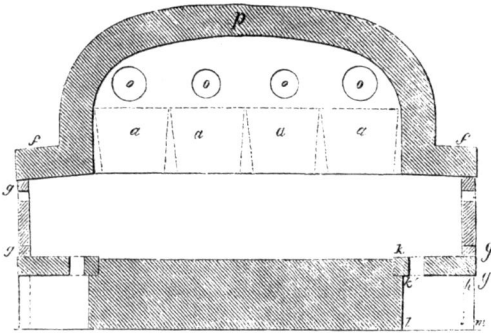

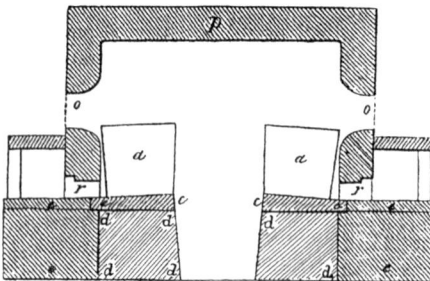

Plan of eight-pot furnace; section A-B through the center of the firebox; vertical section at right angles through the center of pots and holes. *From Georges Bontemps, Guide du Verrier, 1868.*

elder Raybold died (after "ripening for the tomb" for almost a year from an illness that left him bedridden and unable to speak), his obituary surmised that "his loss will be sensibly felt in the business department" of the glass company.[47] His son Walter was superintendent after James's death, with Luther Washburn as overseer in 1870 and William Greaves in 1874.[48] Clearly, overseer and superintendent were two different roles.

Thomas Gaffield was much impressed by how Page turned the company around to produce some of the best window glass in the country. He wrote:

After a good many difficulties & perplexities, which were ingeniously & successfully mastered by the skill, genius & management of Mr. Harrison P. Page, they at length made an article, which, as it was white, as it was new, as it was American, was soon greatly sought after for engraving, for photographers' glass, for windows in houses, & for stores where nice articles of millinery, dry goods &c required a glass which transmitted no color, to display their good qualities.[49]

In 1860, Gaffield visited the factory and provided a thorough description. Shown around by William Harding, at this time an employee, this was Gaffield's first excursion to Berkshire Village, but he would visit a number of times again through the years, so impressed was he by Page and his glassworks. According to Gaffield, at the time of Page & Robbins' purchase, the glassworks had one eight-pot furnace.[50] This would have been the typical furnace of the period as described and illustrated by Georges Bontemps in *Le Guide Du Verrier* (1868), one of the most important texts on glassmaking of the nineteenth century. The furnace was rectangular in plan with an arched roof. Four pots were closely aligned along the interior of each of the long sides, sitting on the *siege*, a ledge raised above a trough about as deep as the pots were tall. The width of the furnace was about three times the diameter of the pots, making the trough about as wide as one pot. The trough extended out the two narrow ends of the furnace forming fireboxes. Fires were laid in the fireboxes and, at appropriate times, were pushed into the furnace along the trough to heat the interior and melt the glass batch contained in each pot. On the exterior long walls, a raised platform, or bench, extended the length of the furnace at the level of the tops of the pots. A circular hole in the furnace wall above each pot provided access for the gatherer.[51] This type of furnace was called a Belgian, German or, later, an American style.

Census records for 1860 show that 341 people lived in Berkshire Village, 47 of whom were employed by the glass factory. Of these workers, one-third (16) were American and 9 were Canadian. There were 7 Irishmen, 6 from present-day Germany (Bavaria and Baden), 4 from Switzerland, 3 Englishmen and 2 Frenchmen. The overseer, furnace man and 3 of the 9 blowers were European.[52]

On April 4, 1861, the *Pittsfield Sun* published an announcement that the company planned to build a second factory building and to buy a sand pit for $2,000 (about $50,000 today).[53] A week later, on April 11, representatives of Page & Robbins issued a testy correction in a competing newspaper as a "precise state of the facts, if the world cares to know them." Calling the previous

English pot, from Hartley & Co., standing nearly five feet high. *From "Examples of Glass-Blowing. By Hartley and Co.,"* Illustrated London News, *September 20, 1851.*

notice "entirely incorrect," they stated that the "Berkshire Glass Company is a very defunct institution." They meant that the name "Berkshire Glass Company" was no longer in use because the factory was now owned by Page & Robbins and went by that name. Furthermore, the company had no need of a sand bed, but the bed referred to in the previous notice had been purchased by Stephen Whipple "on private account" (Whipple, it will be recalled, was the former manager). It was correct, they affirmed, that they were contemplating the construction of new works.[54]

Unfortunately, the Civil War began just one day later, on April 12, resulting in an immediate and precipitous drop in the construction industry. By the start of 1862, the *Sun* reported that Page & Robbins had put its expansion plans on hold, suspending building work begun in 1861.[55] Although glassworkers were exempted from military service, it is likely that the works lost some men to the Union cause.[56]

But the newspaper was not entirely correct, for Page had an "English" furnace built for blowing window-glass cylinders. The English system involved two separate furnaces, one for melting and the other for blowing. Page's melting furnace had ten pots, and the new blowing furnace contained eight *glory holes*, openings in the furnace wall through which the cylinders could be reheated during blowing.[57] By blowing the cylinders away from the pots, the melting furnace could be kept hotter because the gathering hole could be closed while the cylinder was blown at the other furnace. At about twenty-five cubic feet, the pots in an English furnace were twice as large as the American pots. In Europe, this was considered the best way to work, although Page would ultimately disagree.

Page & Harding, 1863–1883

By 1863, the crisis in the building industry precipitated by the Civil War had eased. Page & Robbins was able to begin building again in August of that year. Portions of the old factory were demolished and new buildings erected in their place, including a ten-pot furnace.[58] Page, however, was frustrated. Complaining to Gaffield about the incompetence of workmen and his continuing problem with rusty glass—something he did often—he added to this litany of troubles that he spent everything the factory earned in repairs and improvements.[59] Page said he had stopped using the English furnace because it used too much coal, made the men so hot that they complained and produced glass that was no better than that blown over the pot. He "was so disgusted sometimes that he resolved to sell out" for any price, except that he was reluctant to lose the money he had invested up to now.[60] In September, James Robbins left the firm.[61] William G. Harding, who had worked with the company for about five years, became a partner, and Page hung on to the factory.

William G. Harding

Four months before the Berkshire Glass Works was offered for sale in 1857, a young Bostonian graduated from Williams College in Williamstown, Massachusetts, just fifteen miles from Berkshire Village. William G. Harding, born in Waltham in 1834, had a fine pedigree. A descendant of English colonists, he had been educated at Phillips-Andover Academy and at

Left: William G. Harding, graduation photo, Williams College, 1857. *Williams College Archives and Special Collections, Williamstown, Massachusetts.*

Middle: Page & Harding letterhead. *Williams College Archives and Special Collections, Williamstown, Massachusetts.*

Bottom: Home of William G. Harding in Pittsfield, Massachusetts (demolished). *Pittsfield, Massachusetts: 1761–1911, One Hundred Fiftieth Anniversary Celebration (Pittsfield, Massachusetts: Eagle Printing and Binding Co., 1911). Courtesy of the Berkshire Atheneum, Pittsfield, Massachusetts.*

Specialties.

Roughed & Ribbed Plate.
Rolled Cathedral of various tints.

OFFICE OF

Page, Harding & Co.,

PLATE GLASS IMPORTERS.

Specialties.

Ground & Bent, Dbl. Thick Enameled, Plain & Obscured in neat patterns.

Manufacturers of Berkshire Crystal Glass.

Berkshire, Mass. Feb 14th 1881

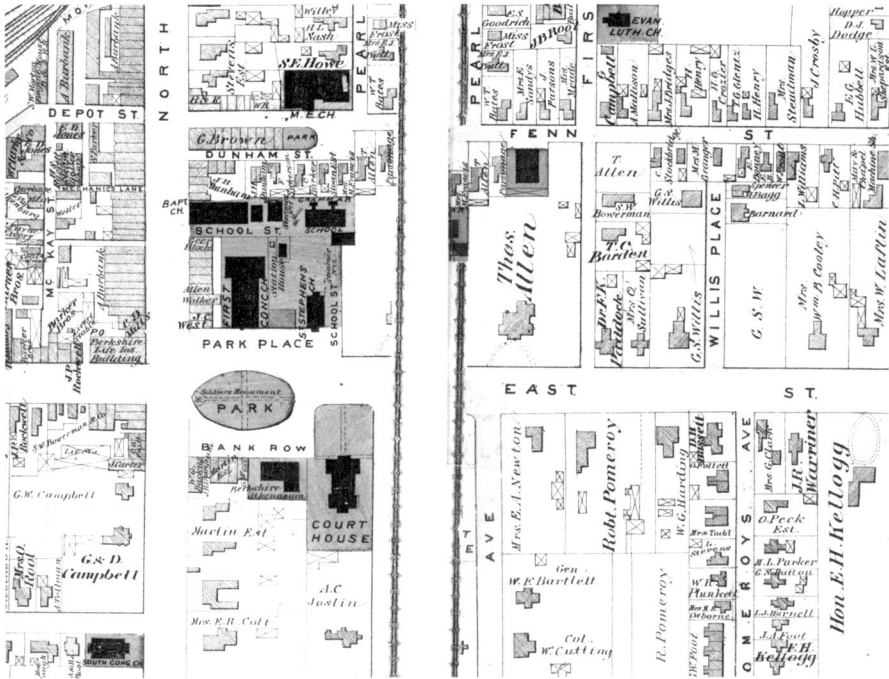

Detail, map of Pittsfield, Massachusetts, showing Harding's house, Thomas Allen's house, First Congregational Church, 1876. *F.W. Beers, County Atlas of Berkshire, Massachusetts (New York: R.T. White & Co., 1876). Courtesy of the Berkshire Atheneum, Pittsfield, Massachusetts.*

Williams had pledged to the Alpha Delta Phi fraternity. Not knowing what else to do after graduation, he returned to the Boston area to teach in a boys' school in Auburndale for a year. By July 1858, he had connected with Harrison Page, who sent him back to Berkshire County to manage the newly purchased glass factory at Berkshire Village.[62] By April 1861, Harding had become a "member of the firm" of Page & Robbins.[63] Two years later, in September 1863, the partnership between Harrison Page and James Robbins was dissolved with Robbins's departure, and the new partnership between Page and Harding became official.[64] Harding's role was to manage the factory, while Page retained control over the techniques of glassmaking. In particular, Harding dealt with employee issues, serving as human resources director, postmaster, negotiator and representative to the trade. He traveled twice to Europe to hire trained glassworkers in Belgium and attended trade organization conferences in the United States.[65]

Harding became a pillar of his community, involved in church and civic activities while also acting as a vigorous promoter of the glassworks. In

1861, he married Nancy (Nannie) Pepoon Campbell (1839–1874), whose prominent family ran one of the local woolen mills. They did not live on the hardscrabble glassworks grounds in Lanesborough among his workers in factory housing and rooming houses but rather in the county seat of Pittsfield, a few miles away. His home, built in 1793 and the oldest in town, was located on East Street at the heart of the city. It had been the old City Hall, which had been moved to this location in 1832. Harding purchased it in 1867.[66] Here he and Nannie raised five children.[67] Events in his life were reported by the local paper, including accidents, the loss of livestock and minstrel shows given at his home.[68]

He served minor roles in local Republican politics. In 1868, he represented Berkshire County as a delegate to the Republican National Convention in Worcester.[69] He was a justice of the peace in Pittsfield for fourteen years and the postmaster at Berkshire Village for forty-three years.[70] He was a founding member of several prominent social clubs in Pittsfield, as well as the Berkshire Historical and Scientific Society. He kept up with his class and fraternity from Williams College, providing reports to alumni newsletters and regularly attending commencement ceremonies.[71]

As a member of the Congregational Church, which he joined in 1861, he served as its deacon from 1900 until 1908.[72] He was deeply involved in activities that promoted the moral edification of local young people as the director of the Young Men's Association, superintendent of and a teacher at the Pontoosuc Sunday School and a member of the Bible Society.[73] For the Berkshire and Columbia (New York) Counties Missionary Society, he was secretary and treasurer.

To promote knowledge and understanding of the glass factory and the manufacture of glass, Harding participated in local fairs and gave tours and lectures of the factory. Showing glass lampshades at the Soldiers Fair in Springfield, Massachusetts, in 1865,[74] where he was awarded $26 ($355 today), he also won for best glass at the Berkshire Agricultural Fair in 1868.[75] His tour and lecture audiences included the Young Men's Association, the Albany (New York) Institute of Science, the Monday Evening Club and the Debating Society of Hinsdale, New York.[76] Perhaps his most important contribution was a long paper on the history of the local glass industry that was published in 1894 in *Collections of the Berkshire Historical and Scientific Society*.

Tragically, his gregarious, man-about-town demeanor diminished with the loss of his wife and two of his five children in 1874. In January, Harding's wife, Nannie, just thirty-four years old, suffered severe burns when she fainted, falling onto a lit kerosene lamp that she carried. She died ten days later.[77] Ten months later, their two youngest children, Hope (b. 1871) and Malcolm (b. 1869), succumbed to scarlet fever within three days of each other.

Harding continued in his work at the glass factory, but his name appeared with less frequency in the social columns of the local newspapers. He never remarried and lived with his three remaining children, none of whom married before his death. He retired in 1902, having spent the previous three years as the treasurer of the Berkshire Co-Operative Glass Company. He died at his home on May 19, 1908, of heart failure, having spent the last years of his life reading and spending time with his children and his church.[78]

* * *

By the end of the Civil War, the glassworks was back on sound financial footing. In September 1865, it was able to purchase 125 additional acres,

The Berkshire Glass Factory's sawmill, purchased by the factory in 1865 (demolished). Photo circa 1910. *Courtesy of David and Sherri Wilson, Berkshire, Massachusetts.*

PRICES CURRENT

OF THE

BERKSHIRE CRYSTAL WINDOW AND PICTURE GLASS,

MANUFACTURED BY

PAGE & HARDING, AT BERKSHIRE, MASS.

PRICES PER 100 FEET.

Sizes.	SINGLE THICK.				THICK.			
	1st Quality.	2d Quality.	3d Quality.	4th Quality.	1st Quality.	2d Quality	3d Quality.	4th Quality.
6x8 to 8x10	$15 00	$14 00	$12 00	$11 00	$22 50	$21 00	$18 00	$16 50
8x11 to 9x13	16 00	15 00	13 00	12 00	24 00	22 50	19 50	18 00
9x14 to 10x15	17 00	16 00	14 00	13 00	25 50	24 00	21 00	19 50
10x16 to 12x18	21 00	19 00	17 00	15 00	31 50	28 50	25 50	22 50
12x19 to 16x24	22 00	20 00	18 00	16 00	33 00	30 00	27 00	24 00
16x26 to 20x24	28 00	24 00	20 00		42 00	36 00	30 00	
20x26 to 20x30	29 00	25 00	21 00		43 50	37 50	31 50	
20x32 to 24x30	33 00	29 00	22 00		49 50	43 50	33 00	
24x32 to 24x36	37 00	34 00	26 00		55 50	51 00	39 00	
26x36 to 30x44	40 00	36 00	27 00		60 00	54 00	40 50	
30x46 to 32x48	44 00	40 00	28 00		66 00	60 00	42 00	
34x48 to 36x48	45 00	42 00	30 00		67 50	63 00	45 00	
Above,	46 00	44 00	32 00		69 00	66 00	48 00	

NUMBER OF LIGHTS IN 50 Feet Boxes.

Size	No.	Size	No.
6x8	150	12x17	35
7x9	114	12x18	34
8x10	90	12x19	32
8x12	75	12x20	30
9x12	67	12x22	27
9x13	62	12x24	25
9x14	57	13x17	33
9x15	53	13x18	31
9x16	50	13x20	28
10x12	60	13x22	25
10x13	55	13x24	23
10x14	51	14x16	32
10x15	48	14x18	29
10x16	45	14x20	26
10x17	42	14x22	23
10x18	40	14x24	21
10x20	36	14x26	20
10x22	33	15x18	27
10x24	30	15x20	24
11x14	47	15x22	22
11x15	44	15x24	20
11x16	41	15x26	18
11x17	39	16x18	25
11x18	37	16x20	23
11x19	34	16x22	20
11x20	33	16x24	19
11x22	30	18x20	20
12x14	43	18x22	19
12x15	40		
12x16	38		

NUMBER OF LIGHTS IN 100 Feet Boxes.

Size	No.	Size	No.
16x26	34	24x40	15
16x28	32	24x42	14
16x30	30	24x44	14
16x32	28	24x48	12
16x34	26	26x30	18
16x36	26	26x32	18
16x38	24	26x34	16
16x40	22	26x36	16
18x24	34	26x38	14
18x26	30	26x40	14
18x28	28	26x42	14
18x30	26	26x44	13
18x32	25	26x46	12
18x34	24	26x48	12
18x36	22	28x32	16
18x38	22	28x34	15
18x40	20	28x36	14
20x24	30	28x40	14
20x26	28	28x42	13
20x28	26	28x44	12
20x30	24	28x48	10
20x32	22	30x36	14
20x34	22	30x38	13
20x36	20	30x40	12
20x38	19	30x42	12
20x40	18	30x44	11
22x26	26	30x46	10
22x28	24	30x48	10
22x30	22	32x40	12
22x32	20	32x42	11
22x34	20	32x44	10
22x35	18	32x46	10
22x38	18	32x48	10
22x40	16	34x40	10
22x42	15	34x42	10
24x28	22	34x44	10
24x30	20	34x48	9
24x32	20	36x44	10
24x34	18	36x46	9
24x36	16	36x48	8
24x38	16	36x50	8

Sizes over 16x24 packed in 100 Feet Boxes.

DISCOUNT. *10 ½ — 30 4 1st Qual* *Sent Prices 1 & 2 d.*   DISCOUNT. *Deu at Factory*

The BERKSHIRE CRYSTAL GLASS is manufactured from the celebrated White Sand found in the vicinity of the Works; and, from its clear light color, and brilliant surface, is superior to any other for Engravings, Daguerreotypes, Photographs, &c.

We manufacture Glass Shades from the same material, Round, Oval, and Square, which are unsurpassed for color and beauty.

Orders addressed to us, at Berkshire or Boston, will have prompt attention.

PAGE & HARDING,

191 STATE STREET.

BOSTON, JUNE, 1864.

"Prices Current of the Berkshire Crystal Window and Picture Glass," June 1864. *Courtesy of Charles Flint.*

The History of the Berkshire Glass Works

View of cylinders at the Berkshire Glass Works, circa 1860s–70s. In the foreground, four cylinders rest on a horse, awaiting capping. Behind them are ranks of cylinders standing on end. *Courtesy of the Berkshire Historical Society, Pittsfield, Massachusetts.*`

View of the factory interior, Berkshire Glass Works, circa 1860s–70s. The man sits on the edge of a barrow used for mixing batch. The wheeled structure in front of him is a carriage used to move a glass pot. In the background against the wall are cylinders. Most have been capped except for a group to the right of center. *From Frances Martin,* Lanesborough, Massachusetts: Story of a Wilderness Settlement, 1765–1965 *(Pittsfield, MA: Eagle Printing, 1965).*

along with a sawmill, a wood shop, a house and water rights from Stephen Whipple.[79] A railroad depot on the Pittsfield & North Adams Line was built for the glassworks in 1866.[80]

The factory employed 130 people by this time but apparently needed even more.[81] Page had changed his mind about his English furnace, deciding to use it, and spent seven weeks in England hiring new craftsmen in late 1866 to work it.[82] Upon his return, it was reported that the factory now had 140 hands, and more than half of the blowers were English.[83] A little more than a year later, it was producing $20,000 to $25,000 (about $272,000 to $340,000) of glass per month.[84] From this period until the mid-1880s, the business of the glassworks was generally secure, although there were months when work was shut down by summer weather (which was typical for glass houses) or by strikes by employees for better wages.

Problems with workers were a common complaint among most glass factory owners, and Berkshire Glass Works was no exception.[85] Although the majority of its workforce was American (according to census records), there was a preoccupation with the behavior of the immigrant workforce, perhaps because of the investment made in getting them to the United States and because they were often the most skilled workers (blowers and gatherers). In early 1863, a worker reported that Page was unhappy with his eight new English workers because upon arriving in Berkshire County—Page having paid their passage—two had gone to work for the Cheshire Glass Works, which presumably had not invested any energy or capital in bringing them to the United States. The rest "became intemperate, discontented, spent their wages, run into debt, left their employers, gone to other factories & some just returned to England."[86] Jealous American workers were unwelcoming, and according to a disgruntled employee, Berkshire Village itself had little to recommend it as a nice place to live, lacking social activities, churches and good schools.[87] Drunkenness continued to be an issue: in 1872, an inebriated factory employee drowned in the reservoir after jumping off the train.[88] Page continued to have difficulty with his English workers, in 1874 citing their frequent intoxication and refusal to blow out entire pots, which, being larger than those in the American-style furnaces, took longer to empty.[89]

By the end of the decade following the Civil War, the Berkshire Glass Works had begun to expand its products from only clear blown window glass. By 1865, it had started blowing colored window glass, and in late 1869, it started to produce plate glass.[90] Plate glass differed from window glass in that it was poured onto an iron table (cast) and rolled, rather than

Rolling cathedral glass. Notice that the iron table is on rails for ease of moving the flattened sheet from the furnace, where it was poured, to the annealing oven. *From* Harper's New Monthly Magazine, *"A Piece of Glass," July 1889.*

blown. The rolling process had been patented in England by James Hartley in 1847.[91] The first plate glass made in the United States had been rolled at the Cheshire Plate Glass Factory, just a few miles away, but that company was in and out of business with some frequency. The production of plate glass was potentially lucrative, since it expanded the number and type of architectural applications of glass from windows to roofs and floors. Plate glass production in the Berkshires was important to the growth of the industry in the United States because it was the first place that John B. Ford (1811–1903), founder of Pittsburgh Plate Glass and Libbey Owens Ford Glass, tried his hand at rolling glass.[92]

The factory ordered a cast-iron bed (or table) on which to roll plate glass (and eventually colored cathedral glass). The massive slab of iron, five and a half inches thick, was three and a half feet wide and forty feet long. It weighed three and a half tons.[93] Before pouring molten glass on the table, it was heated by building a fire below it to prevent the table from cracking from thermal shock, which would happen at the Lenox Plate Glass Factory down the road in 1873.[94]

Morris Schaff, circa 1930s. Proceedings of the Massachusetts Historical Society *64 (March 1932). Courtesy of the Berkshire Atheneum, Pittsfield, Massachusetts.*

In 1872, the firm hired Morris Schaff (1840–1929) as superintendent of the works. Schaff, born in Ohio, was a military man, having graduated from West Point in 1862 and gone straight into the Civil War in the Ordnance Department, where he earned the rank of captain in May 1864. At the end of the war, he was stationed at the Watertown (Massachusetts) Arsenal from January 1865 until September 1866, when he was transferred to Mobile, Alabama, to the Mount Vernon Arsenal, of which he was in command. While stationed there, Schaff fatally shot a Confederate officer named Frederick B. Shepherd in October 1867. Schaff was court-martialed, found guilty of murder and sent to Fort Pulaski in Savannah, Georgia. In January 1868, the charges were reduced to assault with intent to kill, and he was fined $300 and sentenced to six months in prison.[95] He was released from Fort Pulaski in June 1868 and returned to Watertown. Later that year, on August 8, Schaff married Alice, the daughter of Harrison Page, whom he probably had met during his first assignment to the arsenal. The couple spent almost two years at the arsenal in Rock Island, Illinois, until Schaff resigned from the army at the end of 1871.[96] His father-in-law quickly hired him, and he and Alice moved to Pittsfield. Although he knew nothing of making glass, he did speak

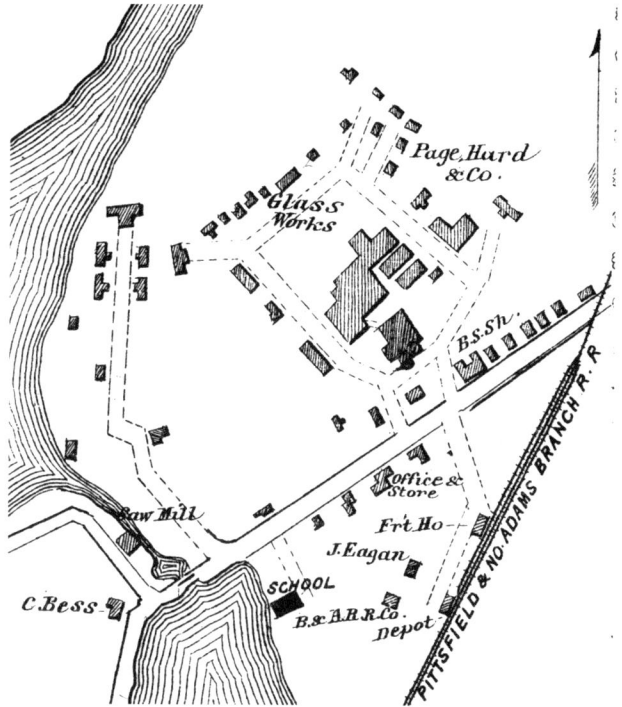

Map of Berkshire Village, 1876. *F.W. Beers*, County Atlas of Berkshire, Massachusetts *(New York: R. T. White & Co., 1876). Courtesy of the Berkshire Atheneum, Pittsfield, Massachusetts.*

French, which enabled him to communicate with the Belgian workers at the factory.[97] He was also knowledgeable about geology and mineralogy, which may have been useful regarding issues of clay and sand.[98] Under Schaff's superintendence, the factory built another glass house.[99] Long after the factory closed, a former employee, Joseph Carrow, complained that Schaff knew nothing about glassmaking and had gotten his job simply through nepotism. Carrow maintained that the ultimate failure of the firm was due to Schaff's mismanagement, but the problem was far more complicated.[100]

FUEL

At Berkshire Glassworks, batch was melted at an average temperature of about 3200 degrees F and then kept at about 2000 degrees for working until the pot was empty.[101] Georges Bontemps wrote that one-third of the cost of running a glass factory is spent on fuel alone (not including shipping) and that since fuel weighed two to three times more than the finished glass, it made economic sense to be close to the source of fuel.[102] One of the greatest impediments to the success of glassmaking in America prior to the turn of the twentieth century was the difficulty in obtaining inexpensive fuel to fire the furnaces.[103] From the time of the first glass house in America at Jamestown, Virginia, in 1608 until the beginning of the third quarter of the nineteenth century, that fuel was wood. The close availability of wood was far more important to the siting of a glass house than was a source of sand. Many glass houses failed when their local forests had been completely consumed in their furnaces. In Berkshire County, both sand and forests were plentiful, but Berkshire Glass Company owner William Harding understood the problem and foresaw the demise of the glass industry in the area. In 1871, Harding referred to the "want of cheap fuel as the great drawback to this Birmingham of America."[104] He was quoted as saying, "If coal-mines could be found in Berkshire, or if her peat-beds could be economically worked, and the peat made *sufficiently dry*, it would soon be the greatest glass-producing country in the world."[105]

Until the mid-1870s, the furnaces of the Berkshire Glass Works were fired first with wood and then later with a combination of coal and wood.[106] In order to burn efficiently, the wood had to be dried, so one of the buildings on site contained wood-drying ovens. This building caught fire one night in 1874, and the factory workers rallied to extinguish the blaze. One man, Ernest Barnard, "regardless of flame and smoke, stood boldly at the oven door, dashing in the water as fast as the buckets could be passed to him." Another, Irishman Martin Carty, dashed to the roof and poured salt down the chimney to smother the flames.[107]

In 1874, the company began making gas for use in its blast furnaces by burning coal oil or petroleum (benzene, or naphtha) to produce the gas.[108] A

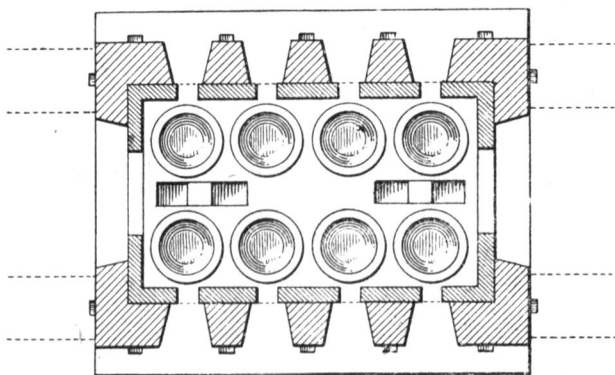

Plan of an eight-pot gas furnace similar to that found at Berkshire Glass Works. *From Percival Marson*, Pitman's Common Commodities in Industries: Glass and Glass Manufacture *(London: Sir Isaac Pitman & Sons, 1918).*

Right: Ruins of the firebox at Berkshire Glass Works. *William J. Patriquin.*

Below: View of the siege at Berkshire Glass Works. On the right is where one pot rested; its shards are visible behind the ring of stones (which is not original or part of the furnace). At the center rear is the grate. *William J. Patriquin.*

contemporary explanation of how "sixty barrels of naphtha a month" was converted into furnace heat quipped, "Steam alone explodes, and naphtha alone explodes, and mixing them, one would suppose that they ought to explode together. But it is found that the two live lovingly together, and go into the furnace hand in hand, and furnish a heat of a great intensity, free

Fire insurance map and view of the Berkshire Glass Works, 1886. See Appendix III for a description of the buildings. *Warshaw Collection of Business Americana—Insurance, Archives Center, National Museum of American History, Behring Center, Smithsonian Institution.*

from some of the objections of coal."[109] In March 1875, Page described to Thomas Gaffield how he was using these materials to heat the furnaces:

> [Page] *conducts the oil from a tank outside of the factory in an inch pipe into the furnace. He also conducts super-heated steam in a large pipe which surrounds the oil pipe until within a short distance from the furnace, where the steam pipe branches off, runs into the furnace in an iron pipe covered with clay & is then returned greatly heated, conducted back to a pipe within about 20 inches of the furnace where the oil & steam unite & form a gas which passes into the furnace, making a great heat & an even heat which can be regulated of course by stopcocks which shall control the amount of the steam and oil used.*[110]

This describes the operation of a Siemens regenerative glass furnace, invented in the early 1860s and patented in the United States in 1872. Like European glass houses that were already utilizing Siemens's invention, Page & Harding found gas to be a superior fuel. It was far more efficient than wood or coal. Gas also maintained the furnace temperature more evenly and eliminated cold spots that made the molten glass ropey or stringy. Because gas burns hotter than wood, the time required to melt a batch was reduced by an hour. Since the gas was created in a separate building, there was no ash or soot in the furnaces that could contaminate the molten glass. So on many counts, the end product was improved.[111] In 1886, the factory had completed its switch to gas as a fuel, and it built two new furnaces.[112] A fire insurance map issued that year shows a gas house located at the center of the factory (see Appendix III for the text on the reverse of the map describing each building). It is a relatively small building in comparison to the glass houses, and one can see how a savings could be had in manpower alone, given the large layout of the factory.

POTS

The clay pots in which the powdered constituents of glass are melted are the most important equipment in the glass house. Their failure by breaking cost a factory much money in lost batch, lost time in removing and replacing the pots in the furnace and possible contamination of other pots in the same furnace.[113]

Berkshire's American pots in 1860 had a capacity of about twelve cubic feet, or 1,100 pounds of batch, out of which about 700 pounds of sheet glass was produced. Approximately 400 pounds from each pot was lost through

Above: View of the west side of the brick pot house (building 7 on the fire insurance map), circa 1940s. *Courtesy of the* Berkshire Eagle, *Pittsfield, Massachusetts.*

Left: Scene in the pot house, Berkshire Glass Works, circa 1890s. *Courtesy of the Berkshire Historical Society, Pittsfield, Massachusetts.*

volatilization, waste from the blowpipes and from capping and breakage during flattening, cutting and packing.[114] In addition, about 3,300 pounds was lost per year because of breakage of the pots.[115] The factory made between seventeen and nineteen "melts" in a month.[116] Each blower made about 70 cylinders of single-thick glass per melt. A single-thick cylinder started with a gather weighing about 11 or 12 pounds.[117] The gather for double-thick glass weighed 25 pounds. The cost to manufacture glass was four dollars per 100 pounds.[118]

In order to ensure quality, most glass factories made their own pots, and Berkshire Glass Works was no exception. Before the Civil War, most clay was imported from Stourbridge, England, home of both glass and ceramics works, but the war disrupted its availability. In 1864, Thomas Gaffield recorded that, if it were available, Stourbridge clay would be selling for four cents per pound, while clay from the Cheltenham and LaClede-Christy mines around St. Louis, Missouri, went for less than two cents per pound.[119] Gaffield's supplier sent him chemical analyses of the three types, and Gaffield was pleased to report that the Missouri clays were as good as, if not even better than, the English material.[120] By 1869, the Berkshire Works was using Cheltenham clay, as were other glass factories around the country.[121] Interestingly, the Cheltenham clay beds in Missouri were owned by Thomas Allen (1813–1882).[122] Allen had been born to a family prominent in the history of Pittsfield and had moved to Missouri in 1842. He was a Missouri state senator from 1850 to 1854 and a federal representative from 1881 to 1882. More importantly, he was the "Railroad King of Missouri," president of the St. Louis, Iron Mountain & Southern Railway, on which Cheltenham was the first stop. He kept his hometown ties active, donating a library (where William Harding sat on the board) to the city in 1876. Allen summered in a large house on East Street in Pittsfield across the street from Harding's home and was undoubtedly acquainted with Harding. He is buried in the Pittsfield Cemetery.[123]

Like most other glassworks, at Berkshire the clay was kneaded by foot, being thoroughly trodden once a day for a month.[124] The raw clay was mixed with *grog*, old potsherds that were ground fine and gave the moist clay more strength and stability. The pots were then hand formed, built up by the coil method—a process that took three months—and left to dry for an additional three months at Berkshire. When a pot was thoroughly dry, it was less likely to break in the furnace. The dried pot was placed in the pot arch of the furnace to temper for about a week—an important step in the process—and then moved to the glass furnace, where it was heated to the temperature for the glass melt. It was then charged, or filled, with batch for melting. Molten

Treading clay for pots. The ceramic rings that floated on top of the metal are in the lower right corner. *From* Harper's New Monthly Magazine, *"A Piece of Glass," July 1889.*

Horse-drawn wheel for crushing potsherds into grog at Berkshire Glass Works (no date). *From Frances Martin,* Lanesborough, Massachusetts: Story of a Wilderness Settlement, 1765–1965 *(Pittsfield, MA: Eagle Printing, 1965).*

The History of the Berkshire Glass Works

Right: Building a pot. *From* Harper's New Monthly Magazine, "*A Piece of Glass,*" *July 1889*.

Below: Joseph Carrow, a former blower, in the remains of the pot house at Berkshire Glass Works, circa 1940s. In front of him is the round wooden bat on which the pot was built. He holds a caliper used to measure the thickness of the walls of the pot and a square. *Courtesy of the* Berkshire Eagle, *Pittsfield, Massachusetts.*

glass is highly corrosive; pots typically last only four to six weeks. In 1861, the factory used sixty-two pots over a working period of forty-five weeks, "or about eight pots to a man, which is considered good work," meaning each pot lasted an average of five to six weeks.[125] Ceramic rings shaped like yokes for livestock rested on the surface of the molten glass to collect the detritus that rose to the top during melting.[126]

* * *

By the mid-1870s, the factory owned seventy tenements in Berkshire Village, which it rented to the employees. But the population of the village was increasing rapidly, and more homes were needed.[127] The largest influx of immigrants was from Belgium, where much of the world's supply of window glass was made.[128] The 1880 U.S. Census listed twenty Belgians working in the glass factory; six were gatherers, three were blowers, one was a cutter, one a melter and the rest were laborers.[129] Many Belgian glassmakers were brought to the United States by glass factory owners because they were highly skilled, having been trained as apprentices from childhood, and because they did not expect the same level of payment as Americans, Englishmen or Germans. The majority of the men earned

One-dollar scrip.
Collection of William J. Patriquin.

48

Twenty-five-cent scrip. *Collection of William J. Patriquin.*

an average of $3.50 a day in piecework (glass blowers and gatherers were typically paid by the number of cylinders they produced). The factory was open three hundred days in 1865.[130]

Like other industries in the mid- to late nineteenth century and through the mid-twentieth century, in the 1870s the Berkshire Glass Works paid its employees in scrip instead of federal currency, to be exchanged for goods at the store in Berkshire Village that was owned by the company.[131] Called "white money" or "white backs" it was issued in denominations of $0.25, $1.00 and $5.00.[132] Printed by Hatch Lithography Co. in New York in both green and black ink, it featured Lady Liberty on the front. In the lower right corner of the front of the $1.00 note was a winsome generic image of cows being herded (or harassed) by a dog on a road lined with telegraph wires. There is a factory in the distance, but it is not a glass factory. On the reverse of the bills, ornate, lacy frames encircled the denomination, which flanked an image of a miner with a pickaxe on the $1.00 note and of a train pulling coal cars on the $0.25 note. The scrip looked like United States currency but was not redeemable at a bank. William Harding would exchange the scrip at his home in Pittsfield on Mondays "at a discount. In order to get square with the discount he charged the people of Berkshire more for the goods sold them than sold elsewhere.

That is the way business was done in Berkshire in those days."[133] Businesses in Pittsfield would accept it as legal tender at 5 percent less than its face value.[134]

In 1873, it was reported that Page & Harding had three factories in Berkshire making plate glass; the story probably meant three *furnaces*. Like its window glass, the company's rough plate glass was considered "superior" and was in great demand for roofing and flooring.[135] In addition, it had added sandblasted glass to the product line in the Boston store where it was produced. The sandblast process had been patented three years earlier and quickly caught on as an inexpensive means of frosting glass. The procedure was so novel that Page & Harding had installed windows on the sandblasting room so that its customers could watch the sheets of glass move on a conveyer belt through a "furious tempest" of sand blown by compressed air.[136] Page had also added bent glass to his product line and had built a leer at Berkshire to cool the bent glass.[137]

Although the factory produced $175,000 (about $3.5 million today) worth of goods in 1874, Page had lost some $100,000 (almost $2 million today) on his English furnaces and, not surprisingly, came to the conclusion once again that the English system was inferior to those in which the blower worked over the pot.[138] By September, he was running three blast furnaces of ten pots each on the American or Belgian plan.[139]

Nevertheless, in early 1878, the company declared bankruptcy.[140] It stated that it had about 1,500 acres of land and thirty to forty houses for the

View of the Berkshire Glass Works, circa 1875–80. *Courtesy of Leo and Ann Sondrini.*

workers and their families (rather than the seventy reported to the newspaper a few years earlier). Page & Harding had $120,000 ($2.67 million today) in liabilities, including a mortgage on the factory for $25,000 ($557,000 today). It was forced to give additional mortgages on real estate and chattel (goods and equipment) in order to raise the money to satisfy its creditors, who included the Boston & Albany Railroad. Gaffield, who represented one of the creditors and so was present for the many meetings held to settle the company's problems, nevertheless felt sympathy for Page:

> *It is doubtful if not used for a glass factory whether* [the property] *would sell for anything above the mortgage.* [Page] *has probably expended upon it 200 to 300,000 dollars.*
>
> *Mr. Page has worked very hard for twenty years, in season & out of season, to advance the character of his work, & has made some very nice white sheet glass. He has also made rolled cathedral & enamelled* [sic] *glass. But the decline in building & in prices, & the great expenses of manufacturing & selling, have crippled him, as many others in the same line & other lines of business have been crippled during the last four years. I sincerely sympathize with him in his misfortune & hope that better is yet in store for him…I sincerely hope he may live long enough to make up for some of the losses of their last few years.*[141]

By April, the debts were paid off for twenty-five cents on the dollar. Page & Harding came out of bankruptcy in June, and the furnaces were restarted in August.[142] But from this point on the company was financed by real estate and chattel mortgages renewed as needed, which was often. Despite the bankruptcy and mortgages, however, it remained a respected firm and was considered busy and prosperous, if not liquid and profit-making.[143] In 1879, glass from Berkshire was being shipped around the world, to New Zealand, the Sandwich Islands, South Africa and South America.[144] Late in the year, three out of five of its furnaces were in production.[145] The company also began to make ribbed plate for glass roofing and remained the only makers of this type of glass in the United States until 1885.[146] In 1880, it was running two ten-pot furnaces and making two hundred boxes of glass per day, earning an estimated $150,000 per year ($3,260,000 today).[147] The factory produced ten thousand square feet of plate glass per week, totaling fifteen tons.[148] The factory employed two hundred men, more than ever before, and by 1886, more than five hundred people lived in Berkshire Village.[149]

BERKSHIRE GLASS COMPANY, 1883–1899

In mid-1883, Page & Harding separated the company into two different firms, leaving the retail store in Boston as Page & Harding and reincorporating the Lanesborough factory as the Berkshire Glass Company under the laws of the State of Maine.[150] The glass company was still admired for the quality of its window glass and the many colors of cathedral glass.[151]

In 1885, Page & Harding took over the Lenox Plate Glass Works under lease and "[took] hold as if they meant business."[152] This factory had been making plate glass since the 1850s, and Page & Harding had been selling the glass in its Boston store since at least 1858 (then as Page & Robbins), the same year the company had purchased the Berkshire Glass Works.[153] The Lenox works had failed in 1872, and subsequent attempts to restart it had been fruitless.[154] Page & Harding had success with it, however, providing twelve thousand square feet of three-quarter-inch-thick plate glass for the new State Library in Austin, Texas, in 1887, as well as rough plate glass for the expansion of the Metropolitan Museum of Art in New York the following year.[155] The firm also provided plate glass to cover almost two acres for the train sheds of the Boston & Lowell and Eastern Railroads in Boston.[156]

LIFE IN THE VILLAGE

Berkshire Village was a very small place. The immediate area around the glass factory—from what today is Summer Street to the Union Chapel and from the railroad to the reservoir—is barely one-quarter of a mile

Top: This view is toward the factory, with smoke coming from the furnace chimneys, circa 1893. The building on the right is the second school. In the distance at center, the roof of the store can be seen through the trees. The building on the left is a residence. *Bottom*: Same view today. *Top: From* Picturesque Berkshire, Part I: North *(Northampton, MA: Picturesque Publishing Co., 1893). Bottom: William J. Patriquin.*

Above: State Road looking north to Berkshire Village, circa 1893. *From* Picturesque Berkshire, Part I: North *(Northampton, MA: Picturesque Publishing Co., 1893).*

Left: *Top*: View of Main Street North. *Bottom*: Present-day view from the same location. *Top: Courtesy of David and Sherri Wilson, Berkshire, Massachusetts. Bottom: William J. Patriquin.*

by half a mile. The larger area encompassing farmland is about three miles square. In 1850, before the glass factory was built, it was a farming community with a population of 225, half of whom were children.[157] In addition to farming, there was a small group of men who worked with wood: carpenters, a sawyer, a lumberman and a wagon maker. The only resident remotely connected to the Berkshire Glass Works was a sand agent, William Fuller. There was no school and no church, although there was a store. Most of the residents were born in America, with a small contingent of Irish immigrants. There was a sizable population of African Americans living in the area known as the Gulf, totaling about 25 percent of the population.[158] This area was along State Route 8 and Gulf Road, near the northern entrance of the present Berkshire Mall. Massachusetts was the home of abolitionism, and all were presumably free. Most were farmers; none were servants.

Ten years later, the population had grown by more than half with the opening of the factory. The number of farmers was halved, although one, Harvey Owens, won a prize for the fertility of his farmlands, which he credited to the ashes from the glassworks that he had spread over his fields.[159] The small lumbering business changed to "wood choppers" who probably harvested timber and cut it up to burn in the glass furnaces. Other trades arose to support the factory, like teamsters and blacksmiths.

The life of a glassworker was physically very hard. Work shifts were often twelve hours long, six days a week, and when the furnaces were in production—typically for ten months of the year, closing in the summer—they were manned around the clock. The furnace buildings were extraordinarily hot. Blowing glass takes a massive amount of lung power, and handling hot glass on the end of a blowpipe requires great strength. Many men—those designated as "laborers"—ran around moving things from building to building. Pay rates between classes of workers varied widely. In 1888, it was reported that an experienced blower could make $400 a month ($9,300 today), while an unskilled laborer was paid only $20 to $30 ($465 to $700) in the same period of time.[160]

Traditionally, around the world children formed an important portion of the glass industry's workforce, especially in factories that made glassware. It appears, however, that few if any under the age of fifteen were employed at the Berkshire Glass Works. Boys worked in the sand beds at Cheshire in 1875.[161] But with the development of the American Window Glass Manufacturers Association (whose members were the factory owners) in the 1880s, American glassworkers' sons were prohibited from apprenticing with

Top: Second school building, circa 1890s. *Bottom*: Second school building today. *Top: Courtesy of the Berkshire Atheneum, Pittsfield, Massachusetts. Bottom: William J. Patriquin.*

their fathers.[162] Women are said to have worked at the factory cutting stencils for enameled glass.

As it became more successful, the factory attracted more immigrants. In 1860, the newcomers were from Canada, Germany, Switzerland, England and France. The African American population decreased from almost 60 to 27; it is not known why this happened.[163] With 170 children (100 of whom attended school) and almost twice as many women as ten years earlier, the town developed a semblance of social refinement, acquiring 2 clergymen, 5 teachers, a store clerk, a butcher and 8 women and girls serving as domestics. A Sunday school opened in 1855, held in the schoolhouse by William Fuller, the sand agent. Fuller remained the Sunday school teacher until his death in 1897. Although there was no church yet, Protestant services were held in the minister's home, over the store, by Reverend Mr. Edson Bonney and his deacon, William Smith, a blacksmith. Attendees included Episcopalians, Congregationalists, Methodists, Baptists, Lutherans "and even some Roman Catholics."[164] When William Harding, a devout Congregationalist, became head of the glassworks in 1858, he reorganized the Sunday school. When Deacon Smith died in 1862, he was replaced by Walter Raybold, whose father was superintendent of the glassworks. He arranged for regular Methodist services to be held in the village, which caused a number of families to convert.[165]

In 1870, as the glass factory reached its golden years, the population swelled to almost 450. The African American population stabilized. Although most were farmers, at least 2 worked at the glass factory as laborers. The factory brought in more English and Irish families to almost double its employee base. The railroad had opened a station in the village. The number of people employed in allied trades dropped, as did the number of farmers, probably because the glass factory paid more. But the number of children swelled to 270, still half of the populace. Despite now having only 1 teacher and no clergymen in residence, religious life was attended to. St. Luke's Episcopal Church in Lanesborough provided a minister for the English families.[166] A window of Berkshire glass adorns the church today. The Pittsfield newspaper reported on weddings and baptisms in the village. One story reported the wedding of a Belgian glass blower whose bride, "Berkshire's fairest daughter," wore "a slate colored Irish poplin, very tastefully made and displaying to advantage her graceful form. A pretty wreath encircled her hat." The couple was driven to their nuptials in "a fine barouche" from Pittsfield, "and it would be hard to find anywhere a handsomer quartet" than the bridal party.[167] In 1871, the second schoolhouse (still standing) was erected, and Sunday services were held there: Catholic Mass in the morning, Sunday school at midday and Protestant services in the evening. Despite the number of Irish immigrants living in the village, there was only one Catholic wedding held there.[168]

During the factory's most successful years in the 1880s, the village's population reached a peak of 478, with a quarter of these employed by the

Top: Union Chapel, 1888; photo circa 1893. *Bottom*: Union Chapel today. *Top: From* Picturesque Berkshire, Part I: North *(Northampton, MA: Picturesque Publishing Co., 1893). Bottom: William J. Patriquin.*

glassworks. Interestingly, this is the only year in which men were unemployed because they were elderly, and the community hosted two "paupers." Although many of the English families stayed in the village, they were supplanted in the factory by Belgians. The African American population dropped from 42 to 19, although the glassworks continued to employ several as laborers. Although there were no resident ministers and still only 1 teacher, a business in women's clothing began, employing 2 people, a dressmaker and a shoemaker.[169]

Frustrated by the inadequate size and furnishings of the school for religious services, a nondenominational congregation was legally organized in 1887, with William Fuller as president, and undertook the building of a chapel. Local families were involved, including Snows, Williamses, Martins, Hineses, Hardings, Reddings and Stevenses. Fundraising had begun during the summer of 1886, led by local women from Berkshire Village and Lanesborough.[170] Land and money were donated by Harvey Chase and his family and by the glassworks. The Boston & Albany Railroad provided the stone for the foundations, which was taken from the abutments of a bridge that spanned the tracks on the property. Designed by local architect H. Neill Wilson from Pittsfield, ground was broken in October 1887 by two of the village's oldest female residents, including Mrs. Harvey Chase, who, "standing…on the verge of the Great Hereafter, look with far-seeing eyes beyond the present, with prayers in their secret hearts for the hundred-fold harvest of the seed sowing on that part of their farm." When the cornerstone was laid by William Fuller, it contained a copper time capsule with four local newspapers (three from Pittsfield and the Springfield *Republican*), lists of contributors and teachers and students at the Sunday school, the bylaws of the congregation and a history of the Sunday school. Although the building contractors were from Pittsfield, local residents worked on the actual building, including Alonzo Hoose, "an aged and much respected black man [who] lived to attend many a Sunday-school and service in the completed chapel from which his funeral was held at last." The Union Chapel was completed in late 1888, dedicated two days after Christmas. Its windows, donated by the Ladies Aid Society of the congregation and by William Harding, were made with colored glass donated by the Berkshire Glass Works.[171]

A vibrant social life had developed in the village by this time. There were two baseball teams, the Clippers and the First Nine, and an orchestra. Dances, socials and fairs were held at the center of the village in front of the store.[172] A Ladies Aid Society was founded in 1887. Amateur dramatic productions were performed to raise money for the chapel.

One notable absence from the lists of residents is the profession of doctor. Presumably, one had to call to Pittsfield or one of the surrounding villages for medical help. This must have been a problem for a working population involved with fire and heavy weights. For example, in 1858, the leg of one Jethro Tucker was crushed by a loaded wagon. One wonders how long Mr. Tucker had to suffer in excruciating pain until the doctor arrived.[173] Similarly, when a bent (part of the structure of the building) fell in one of the factory buildings, three men were badly injured.[174] Less than a month later, one of the French workers fell from a train and suffered head injuries.[175] In more dire circumstances, it took only a day for help to arrive from Boston, as in the case of the death from smallpox in 1889 of the wife of a Belgian glassblower. A representative of the state board of health was sent to investigate the source of the outbreak. The deceased woman had just arrived in Berkshire Village from Europe and had traveled in steerage. Among the passengers was a child with smallpox. Although the woman had been vaccinated, she contracted the disease anyway, causing a general scare in the village that resulted in the closure of the village's two schools. The entire family was moved from their house in the village, which was located among the factory's tenements, and made to stay in tents in Charles Best's upper pasture some distance away from the rest of the populace. She died within a week of her arrival and is rumored to have been buried in the pasture, away from the village.[176]

In 1900, as the factory struggled, the population of the village fell by almost 50 percent.[177] From a high of 111, the glassworks now employed only 27 residents. An additional 34 people were listed as "day laborers" and it is not clear whether they worked for the glassworks or for the 16 farmers left in the community. The Belgians and most of the English, Swiss and Germans moved on, since they were skilled laborers and could find work in the burgeoning glass industry in the natural-gas belt from West Virginia to Indiana.[178] They took their large families with them, and the number of children dropped to an all-time low of 85, with 53 women.

* * *

Success for the Berkshire Glass Company was short-lived. A combination of circumstances that heralded the beginning of enormous success in the American glass industry as a whole signaled the beginning of the end for Berkshire. In the early 1880s, glassworkers in the larger factories of the

Midwest and western Pennsylvania had begun to band together in unions and exercise their demands for better wages and hours through strikes.[179] A frequent complaint among them was that Europeans would work for less money.[180] When a walkout was called, it affected every glass house in the country, and these walkouts became annual events in the late 1880s and early 1890s. Small factories like the Berkshire Glass Company had to abide by the wage settlements made for the industry as a whole, whether they could afford to or not. Glass blowers were paid more than any other glassworkers because of the skill and physical strength required. In 1890, the average glass blower working in Pennsylvania earned $30.85 per day—ten times as much as he had been paid in 1870. In Massachusetts, however, factory owners were permitted by the unions to pay 10 percent less because of their increased expenses, which had the opposite effect than intended—instead of giving the Massachusetts factories like Berkshire Glass an advantage, it just encouraged their skilled workforce to move to greener pastures.[181]

The discovery of natural gas in western Pennsylvania tolled the death knell for eastern glassworks. As early as 1875, this fuel was being used to fire glass furnaces.[182] In 1898, one author wrote that "the direct coal-fired plants of twenty years ago cannot compete, either in quality of product or economy of operation, with the gas fired plants of the present time."[183]

In addition, a new type of furnace, the tank furnace, opened at Jeanette, Pennsylvania, in 1889.[184] The tank furnace, a further adaptation of the Siemens, did away with individual pots, melting the glass in one large vat or tank that was continually charged at one end and gathered or dipped out at the other. It was a brilliant concept that helped revolutionize the glass industry. But to change the furnaces was costly, and Berkshire could not, or would not, do it. In 1890, the number of men employed at Berkshire dropped to 110 from its high of 200 in 1880.[185]

The company stated that the biggest impediment toward making window glass at a price low enough to compete with imported glass was the cost of labor, which was dictated by the union, which "decline[d] to yield anything."[186] In addition to the expense of labor, the cost of shipping coal to Berkshire Village became an enormous obstacle to making glass at a competitive price by the 1880s. Harding complained of this in the mid-1890s.[187] As early as 1876, Page & Harding led a group of local glassmakers to petition the state legislature to regulate rail freight costs for shipping coal from the Hudson River into Berkshire County.[188] To get their glass to market, it traveled the same route in reverse: to ship glass to New York or Boston, it was picked up at the railroad depot in Berkshire Village and taken by train to Hudson,

New York. There it was loaded onto ships or barges in the Hudson River and floated south to New York, where it could then be shipped anywhere in the world. Harding maintained that it was less expensive to ship glass from Belgium than from Berkshire to New York and that the freight rates for that journey were more than 25 percent higher than for any other American glass factory, because there was no competition.[189]

The window and plate glass industry was rapidly expanding and moving toward the natural gas belt in western Pennsylvania, Ohio, and Indiana. The number of window glass factories in the United States doubled between 1880 and 1890, and the number of factories making cathedral glass increased eight-fold.[190] Even though Berkshire was the only window-glass factory in Massachusetts, the value of its products dropped by half between 1880 and 1890, from $854,000 to $431,000, or from 4 percent of the national output to barely 1 percent.[191] The Berkshire Glass Works could not keep up. In 1887, the Mississippi Glass Company of St. Louis, Missouri, became Berkshire's first competitor in the production of ribbed plate and cathedral glass.[192] Berkshire no longer held a monopoly on any type of glass. By 1900, in the production of cathedral glass, Massachusetts now ranked last with 16 pots, far behind Missouri's 120 pots.[193] Without a unique product, Berkshire did not have the resources to be competitive. Compounding the agony was the financial depression caused by the Panic of 1893, which caused an enormous stagnation in the sales of window glass, among all other manufactured goods.[194]

The Berkshire Glass Company filed for bankruptcy in February 1894 and was purchased by Harrison Page's wife at auction for "a little less than $23,000 [slightly less than $600,000 today]," barely enough to settle the mortgage.[195] Morris Schaff had resigned the previous year, and William Harding and Harrison Page were no longer young men.[196] Despite a local claim that the factory was purchasing five tons of sand a day in 1895, another article in 1897 stated that the factory had not been operational for two years.[197] In early 1897, just when the Berkshire Glass Works was trying to restart, the Window Glass Workers Association suffered from a paralyzing internal dispute between the glass blowers and gatherers, who earned the most, and the flatteners and cutters. Manufacturers were unable to negotiate with the association until this problem was worked out. The workers did not return to the factories until early the following year.[198] But then Harrison Page died on July 15, 1898. In August, Harding announced that he and the new manager, Charles Cummings, had no plans to reopen and the employees were looking for new jobs.[199] The factory did run for a month that year, but in March 1899 announced it would close for an indefinite period.[200] Its time as a closely owned corporation was over.

BERKSHIRE CO-OPERATIVE GLASS COMPANY, 1899–1903

By the end of 1899, a group of employees of the Berkshire Glass Works, with additional hands from the Chicago area, decided to restart the factory as a cooperative owned by the employees with $50,000 ($1,340,000 today) in capital. The capital, it was said, came from New York, from a concern that also financed four other glass cooperatives. The president of the Berkshire Co-Operative Glass Company was Conrad Hines, who had spent his life working at the factory as a flattener.[201] The vice-president was William Kelly, a glass blower, and the secretary was J.C. Hines. William G. Harding, now sixty-five years old, was the treasurer. With the celebrated but futile optimism of the underdog, they vowed to go up against the American Glass Company, a combine of 75 percent of the glass factories in the United States that was controlled by the factory owners.[202] A small but enthusiastic contingent of workers established eight cooperatives around the country in the late 1890s. Their goal was to break the stranglehold of the manufacturers' association, which had limited the amount of time its factories would be in production to six or seven months of the year. Berkshire Glass Works had never belonged to the manufacturers' association, which is probably why the workers who reorganized the factory invited William Harding to be an officer of the new cooperative.[203]

Within three weeks of announcing their plans, the officers had started up one ten-pot furnace with sixty employees. An ominous note delayed the opening by a few days: a shipment of coal was held up.[204] But the furnaces started eventually, and the first new cylinders were blown in October 1899.[205]

View of abandoned factory buildings looking northwest, circa 1910. *Courtesy of David and Sherri Wilson, Berkshire, Massachusetts.*

But in 1901, a small notice in the *New York Times* said that the Berkshire Co-Operative had "been idle for several years."[206] In 1903, it was three years delinquent on taxes.[207] By 1904, "owing to competition in the coal producing states," the factory had "almost become a ruin."[208] As Harding had predicted as early as 1871, the lack of fuel had indeed doomed the factory.

Several of the sand mines continued in business into the 1950s or '60s. In 1948, the Berkshire Glass Sand Company was still shipping twenty tons of sand to Tel Aviv, Israel, annually, where it was used to make porcelain teeth and fillings.[209] In 1951, local ownership ended, although the company was still in business in 1957.[210]

In the 1940s, most of the remaining factory buildings were demolished. The ruins of one of the furnaces survive. Chunks of cullet and shards of clear and colored glass can still be found in the ground. The remains of the sand mines are grown over. Many of the houses built for workers still remain today, as does the second school building and the Union Chapel. Descendants of the glassworkers still reside in the area. No glassmaking or sand mining businesses exist anymore.

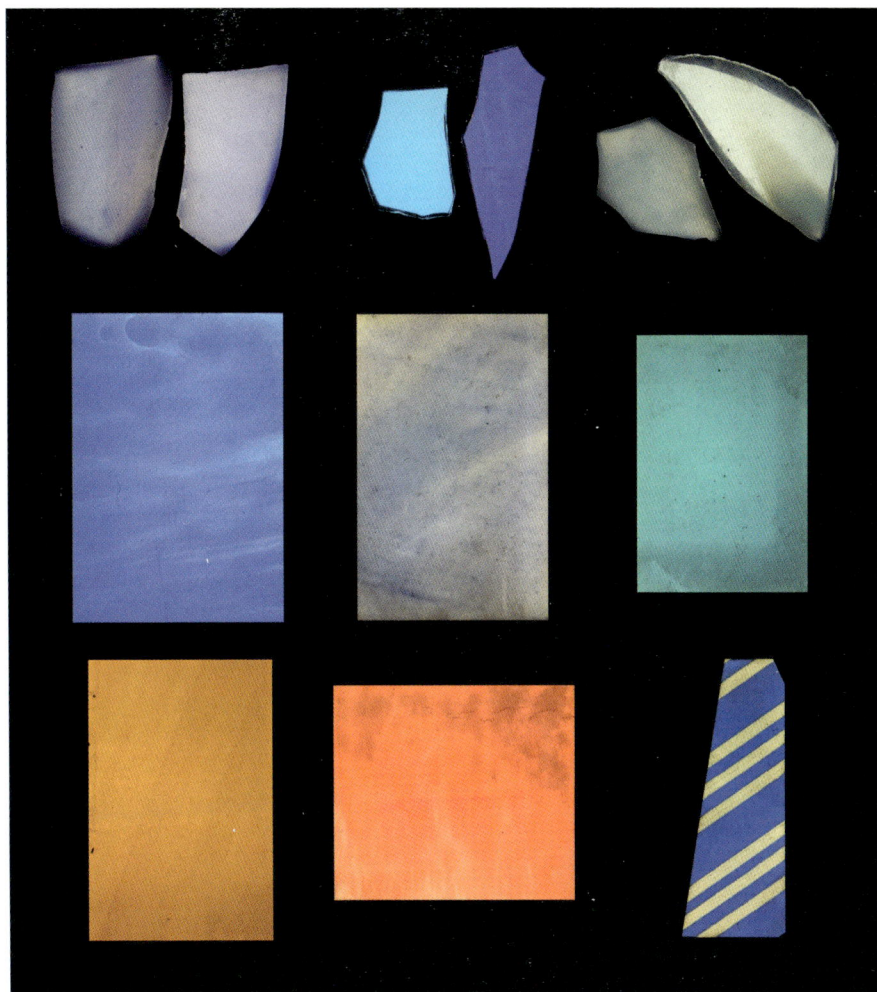

Antique glass from the Berkshire Glass Works. *By row, left to right from the top*: Antique glass from the Berkshire Glass Works: pale pinkish blue shards. *Julie L. Sloan. Collection of William J. Patriquin*; Antique glass from the Berkshire Glass Works: *Left*: copper blue shard. *Right*: mazarine blue shard. *Julie L. Sloan. Collection of William J. Patriquin*; Antique glass from the Berkshire Glass Works: glacial ice blue shards. *Julie L. Sloan. Collection of William J. Patriquin*; Antique glass from the Berkshire Glass Works: medium blue sample. *Julie L. Sloan. Collection of the Berkshire Museum, Pittsfield, Massachusetts*; Antique glass from the Berkshire Glass Works: pale blue. *Julie L. Sloan. Collection of the Berkshire Museum, Pittsfield, Massachusetts*; Antique glass from the Berkshire Glass Works: turquoise sample. *Julie L. Sloan. Collection of the Berkshire Museum, Pittsfield, Massachusetts*; Antique glass from the Berkshire Glass Works: amber. *Julie L. Sloan. Collection of the Berkshire Museum, Pittsfield, Massachusetts*; Antique glass from the Berkshire Glass Works: terra cotta. *Julie L. Sloan. Collection of the Berkshire Museum, Pittsfield, Massachusetts*; Shard of blue flashed glass with wheel engraving, shard found at the Berkshire Glass Works. Red examples have also been found. *Julie L. Sloan. Collection of William J. Patriquin.*

Streaky glass from the Berkshire Glass Works. *By row, left to right from the top*: Sample of red-and-blue streaky. *Julie L. Sloan. Collection of the Berkshire Museum, Pittsfield, Massachusetts*; Sample of red-and-blue streaky in reflected light. *Julie L. Sloan. Collection of the Berkshire Museum, Pittsfield, Massachusetts*; Sample of red-and-blue streaky. *Julie L. Sloan. Collection of the Berkshire Museum, Pittsfield, Massachusetts*; Sample of red-and-blue streaky. *Julie L. Sloan. Collection of the Berkshire Museum, Pittsfield, Massachusetts*; Sample of red-and-clear streaky. *Julie L. Sloan. Collection of the Berkshire Museum, Pittsfield, Massachusetts*; Sample of red streaky. *Julie L. Sloan. Collection of the Berkshire Historical Society, Pittsfield, Massachusetts*; Sample of red-and-clear streaky. *Julie L. Sloan. Collection of the Berkshire Museum, Pittsfield, Massachusetts*; Sample of red-and-yellow streaky. *Julie L. Sloan. Collection of the Berkshire Historical Society, Pittsfield, Massachusetts*; Sample of amber-and-caramel streaky. *William J. Patriquin. Collection of the Berkshire Museum, Pittsfield, Massachusetts.*

Streaky glass from the Berkshire Glass Works. *By row, left to right from the top*: Sample of amber-and-brown streaky. *Julie L. Sloan. Collection of the Berkshire Museum, Pittsfield, Massachusetts*; Sample of amber streaky with bubbly surface texture. *Julie L. Sloan. Collection of the Berkshire Historical Society, Pittsfield, Massachusetts*; Sample of pinkish-brown streaky. *Julie L. Sloan. Collection of the Berkshire Museum, Pittsfield, Massachusetts*; Shard of amber-and-green streaky. *Julie L. Sloan. Collection of William J. Patriquin*; Sample of olive-green streaky. *Julie L. Sloan. Collection of the Berkshire Museum, Pittsfield, Massachusetts*; Shard of deep moss-green-and-clear streaky. *Julie L. Sloan. Collection of William J. Patriquin*; Sample of deep purple-and-clear streaky. *Julie L. Sloan. Collection of the Berkshire Museum, Pittsfield, Massachusetts*; Sample of purple streaky. *Julie L. Sloan. Collection of the Berkshire Museum, Pittsfield, Massachusetts*; Sample of purple streaky in reflected light. *William J. Patriquin. Collection of the Berkshire Museum, Pittsfield, Massachusetts*.

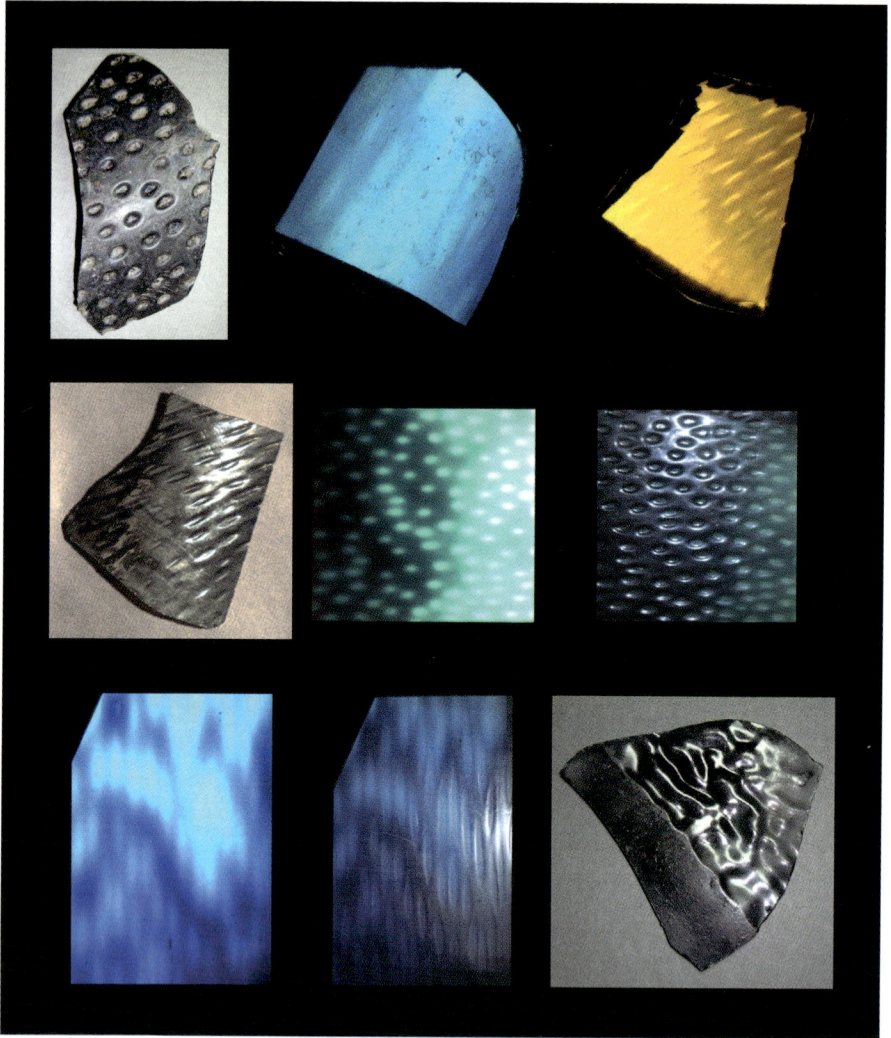

Blown textured glass made at the Berkshire Glass Works. *By row, left to right from the top*: Shard of a venetian-glass cylinder showing the dimpled texture. This piece was not blown out. *Julie L. Sloan. Collection of William J. Patriquin*; Shard of a venetian-glass cylinder showing the dimpled texture. This piece was blown out. Notice that the dimples are very elongated. *Julie L. Sloan. Collection of William J. Patriquin*; Shard of a yellow venetian glass cylinder. *Julie L. Sloan. Collection of William J. Patriquin*; Shard of a yellow venetian glass cylinder in reflected light. *Julie L. Sloan. Collection of William J. Patriquin*; Sample of teal venetian glass. *Julie L. Sloan. Collection of the Berkshire Museum, Pittsfield, Massachusetts*; Sample of teal venetian glass in reflected light. *William J. Patriquin. Collection of the Berkshire Museum, Pittsfield, Massachusetts*; Sample of blue venetian glass. *Julie L. Sloan. Collection of the Berkshire Museum, Pittsfield, Massachusetts*; Sample of blue venetian glass in reflected light. *William J. Patriquin. Collection of the Berkshire Museum, Pittsfield, Massachusetts*; Shard of blown textured glass. *Julie L. Sloan. Collection of William J. Patriquin*.

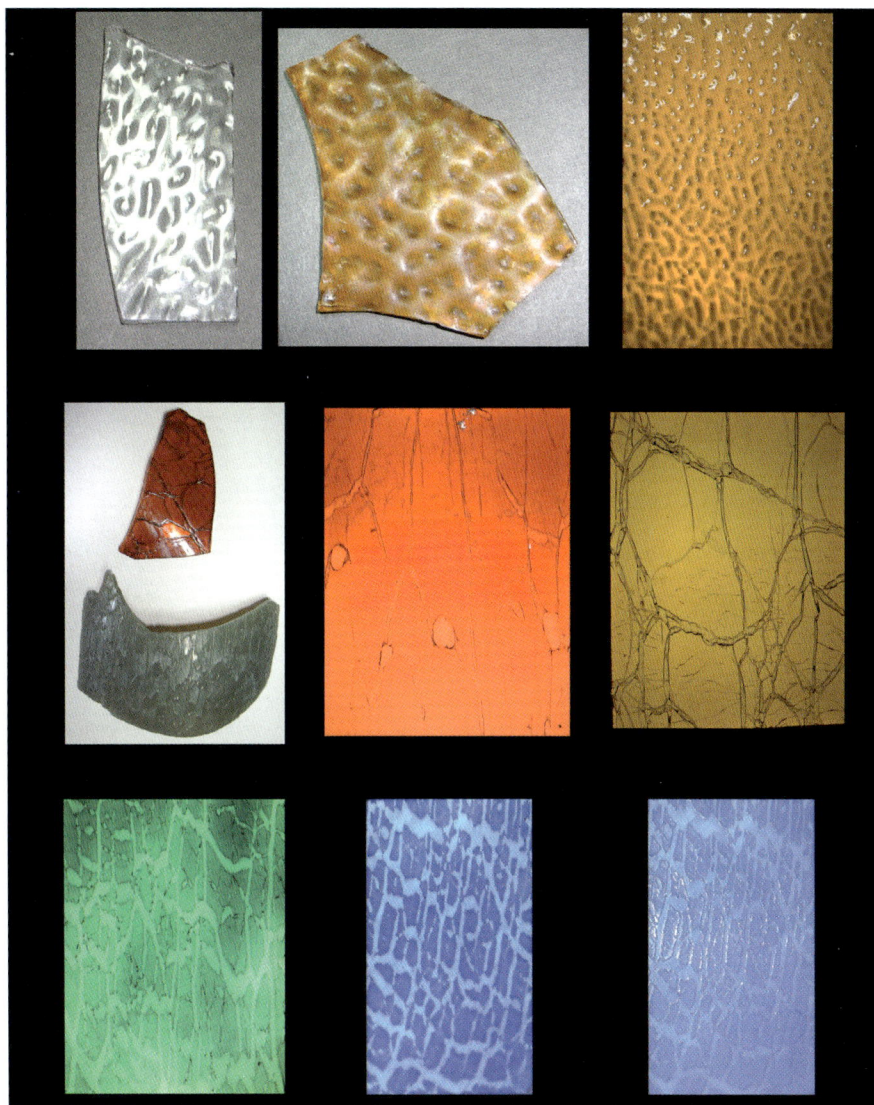

Blown textured glass made at the Berkshire Glass Works. *By row, left to right from the top*: Shard of blown textured glass. *Julie L. Sloan. Collection of William J. Patriquin*; Shard of blown textured glass. *Julie L. Sloan. Collection of William J. Patriquin*; Sample of blown textured glass. *Julie L. Sloan. Collection of the Berkshire Museum, Pittsfield, Massachusetts*; Shards of crackled glass cylinders. *Julie L. Sloan. Collection of William J. Patriquin*; Sample of red crackled glass. *Julie L. Sloan. Collection of the Berkshire Museum, Pittsfield, Massachusetts*; Sample of gold crackled glass. *Julie L. Sloan. Collection of the Berkshire Museum, Pittsfield, Massachusetts*; Sample of green crackled glass. *Julie L. Sloan. Collection of the Berkshire Museum, Pittsfield, Massachusetts*; Sample of blue crackled glass. *Julie L. Sloan. Collection of the Berkshire Museum, Pittsfield, Massachusetts*; Sample of blue crackled glass in reflected light. *Julie L. Sloan. Collection of the Berkshire Museum, Pittsfield, Massachusetts*.

Cathedral glass made at the Berkshire Glass Works. *By row, left to right from the top*: Shards of cathedral glass in reflected light. *Julie L. Sloan. Collection of William J. Patriquin*; Shard of brown cathedral glass. *Julie L. Sloan. Collection of William J. Patriquin*; Sample of dusty pink cathedral glass. *Julie L. Sloan. Collection of the Berkshire Museum, Pittsfield, Massachusetts*; Numbered samples of cathedral glass, dug from the site of the Berkshire Glass Works. *Julie L. Sloan. Collection of William J. Patriquin*; Green cathedral glass from tenement window. *William J. Patriquin*; Pink cathedral glass from tenement window. *William J. Patriquin*; Clear cathedral glass, First Baptist Church, Cheshire, Massachusetts. *William J. Patriquin*; Clear cathedral glass, First Baptist Church, Cheshire, Massachusetts. *William J. Patriquin*; Sample of typical "beef-steak" glass. *William J. Patriquin. Collection of the Berkshire Museum, Pittsfield, Massachusetts.*

Cathedral glass made at the Berkshire Glass Works. *By row, left to right from the top*: Sample of unusual "beef-steak" glass. *Julie L. Sloan. Collection of the Berkshire Museum, Pittsfield, Massachusetts*; Sample of unusual "beef-steak" glass. *Julie L. Sloan. Collection of the Berkshire Museum, Pittsfield, Massachusetts*; Shard of gold ribbed window glass. *Julie L. Sloan. Collection of William J. Patriquin*; Shard of green ribbed window glass. *Julie L. Sloan. Collection of William J. Patriquin*; Shard of ice-blue ribbed window glass. *Julie L. Sloan. Collection of William J. Patriquin*; Sample of clear reeded glass. *Julie L. Sloan. Collection of the Berkshire Museum, Pittsfield, Massachusetts*; Sample of gold reeded glass. *William J. Patriquin. Collection of the Berkshire Museum, Pittsfield, Massachusetts*; Sample of green reeded glass. *Julie L. Sloan. Collection of the Berkshire Museum, Pittsfield, Massachusetts*; Sample of dusty-pink reeded glass. *Julie L. Sloan. Collection of the Berkshire Museum, Pittsfield, Massachusetts*.

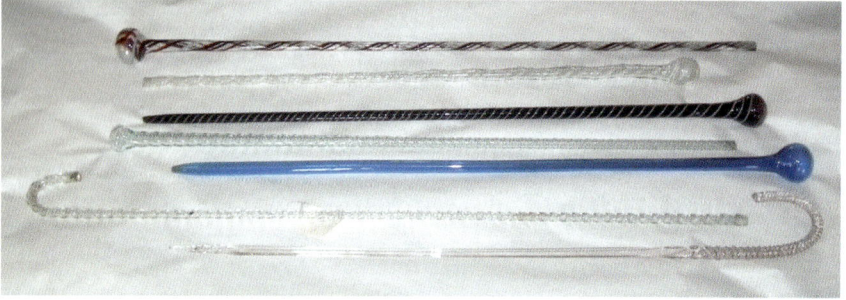

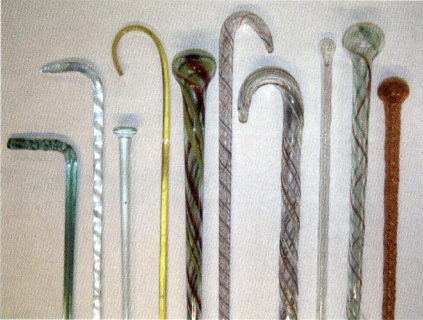

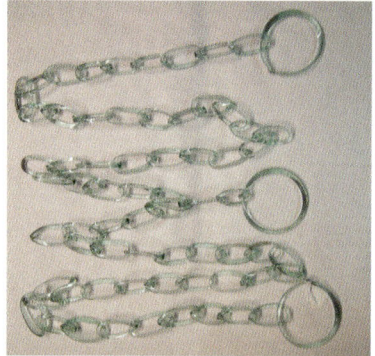

Whimsies made at the Berkshire Glass Works. *By row, left to right from the top*: Glass canes made at the Berkshire Glass Works. *Julie L. Sloan. Collection of the Berkshire Historical Society, Pittsfield, Massachusetts*; Glass canes made at the Berkshire Glass Works. *Julie L. Sloan. Collection of the Berkshire Museum, Pittsfield, Massachusetts*; A glass chain made at the Berkshire Glass Works. *Julie L. Sloan. Collection of the Berkshire Museum, Pittsfield, Massachusetts*; Drinking glasses and a Christmas ornament made at the Berkshire Glass Works. *William J. Patriquin. Collection of Ed and Kim LaMarre*; Witch's ball made at the Berkshire Glass Works. *William J. Patriquin. Collection of Ed and Kim LaMarre.*

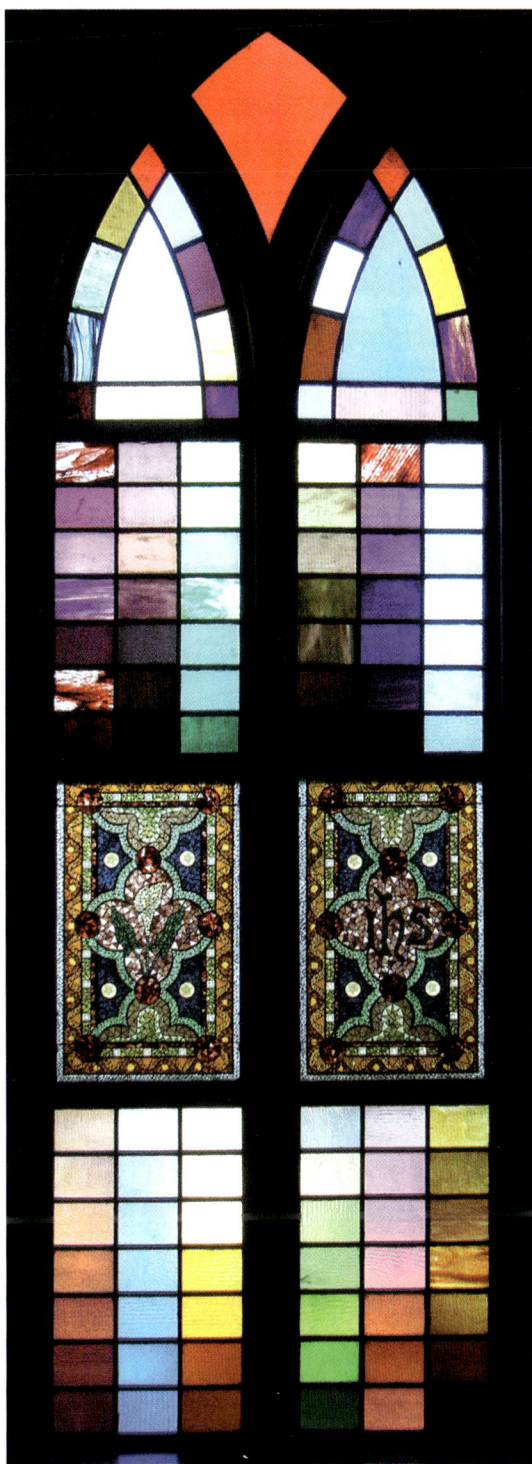

Window, St. Luke's Episcopal
Stone Church, Lanesborough,
Massachusetts, circa 1880. *Julie
L. Sloan.*

Above: Shards of cylinders dug from the Berkshire Glass Works site. *Julie L. Sloan.*

Left: *Top*: Shards of pieces cut for a window, dug from the Berkshire Glass Works site. *Bottom*: Shards of painted glass, dug from the Berkshire Glass Works site. *Julie L. Sloan. Collection of William J. Patriquin.*

Right: Shards of clear enameled glass dug from the Berkshire Glass Works site. *Julie L. Sloan. Collection of William J. Patriquin.*

Bottom: Enameled glass. *By row, left to right from the top*: Red enameled glass from Berkshire Glass Works. *Julie L. Sloan. Collection of the Berkshire Museum, Pittsfield, Massachusetts*; Green enameled glass from Berkshire Glass Works. *Julie L. Sloan. Collection of the Berkshire Museum, Pittsfield, Massachusetts*; Enameled pattern from Chance Brothers, Birmingham, England. *Chance Brothers,* Patterns of Enamelled, Double-Etched, and Stained Enamelled Glass *(Birmingham, 1863)*; Enameled pattern from Chance Brothers, Birmingham, England. *Chance Brothers,* Patterns of Enamelled, Double-Etched, and Stained Enamelled Glass *(Birmingham, 1863)*; Enameled pattern from Carter Brothers. This design is in the bottom center of the photograph of the shards. *Carter Brothers,* Ornamental Glass *(Pittsburgh, n.d.), collection of Julie L. Sloan*; Pattern from Carter Brothers. This design is the second piece from bottom right in the photograph of the shards. *Carter Brothers,* Ornamental Glass *(Pittsburgh, n.d.), collection of Julie L. Sloan*; Enameled pattern from Chance Brothers, Birmingham, England. This design, without the yellow stain, can be seen in two pieces in the center of the photograph of the shards. *Chance Brothers,* Patterns of Enamelled, Double-Etched, and Stained Enamelled Glass *(Birmingham, 1863)*.

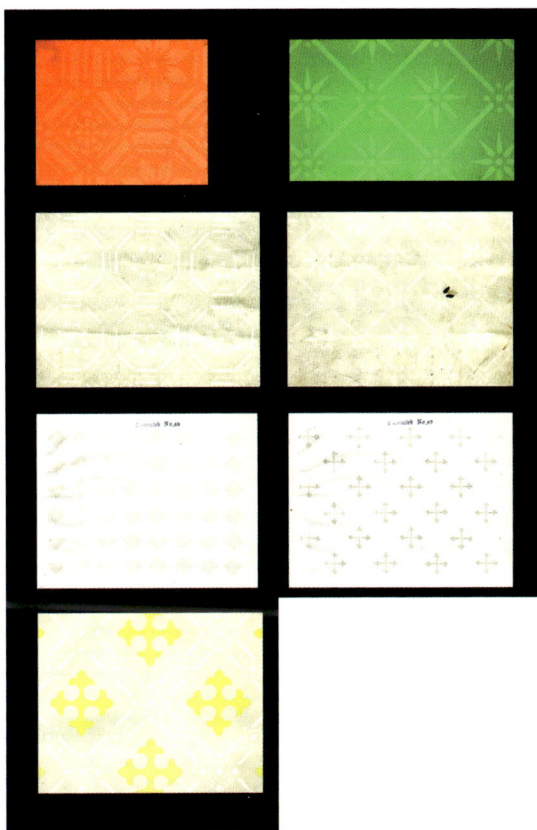

Left: Lantern in the main hall at Chesterwood, the home of Daniel Chester French, Lenox, Massachusetts. This is the same pattern as the green sample. *Julie L. Sloan. Collection of Chesterwood, Lenox, Massachusetts. Courtesy of the National Trust for Historic Preservation.*

Below: Cook, Redding & Co., reading room window, Waterloo Library, 1882, Waterloo, New York, made of cathedral glass made by the Berkshire Glass Works. *Julie L. Sloan*

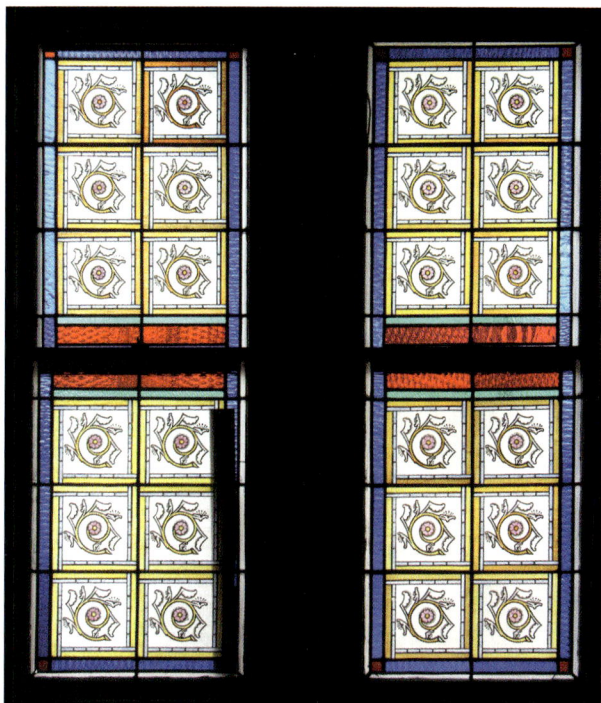

W.J. McPherson & Co., window in parish hall, 1875, Unity Church, North Easton, Massachusetts. *Julie L. Sloan.*

Detail, W.J. McPherson & Co., window in parish hall, 1875, Unity Church, North Easton, Massachusetts, with blue and red venetian glass and clear ribbed glass from the Berkshire Glass Works. *Julie L. Sloan.*

Donald MacDonald,
Charity and Devotion,
1872, St. Ann's
Episcopal Church,
Lowell, Massachusetts.
Julie L. Sloan.

Redding, Baird
& Co., *Moses*,
circa 1883, First
Congregational
Church, Malone,
New York. *Julie L.
Sloan.*

Opposite, top: Pile of shards dug from the site of the Berkshire Glass Works. *Julie L. Sloan.*

Opposite, bottom: The two large chunks of cullet—blue glass on the left and clear on the right—sit on a fragment of a glass melting pot. In the center between them is a pile of Berkshire County sand. Resting on top of the clear cullet is a piece of Cheshire quartzite, which was pulverized to make the sand. In the rear is a sheet of Berkshire County clear plate glass. Hanging from it is a glass chain, a whimsy possibly from the factory. The glass cane is filled with Berkshire County sand. *Courtesy of Charles Flint.*

This page, top: Detail, window made from shards of dusty-pink, gold and blue Berkshire glass from Willimantic Linen Co., Mill Number 4, Windham Textile Museum, Willimantic, Massachusetts. *Julie L. Sloan.*

This page, bottom: Detail, John La Farge, *Manga*, circa 1875, Walter Hunnewell house, Wellesley, Massachusetts, made of Berkshire cathedral glass. *Julie L. Sloan.*

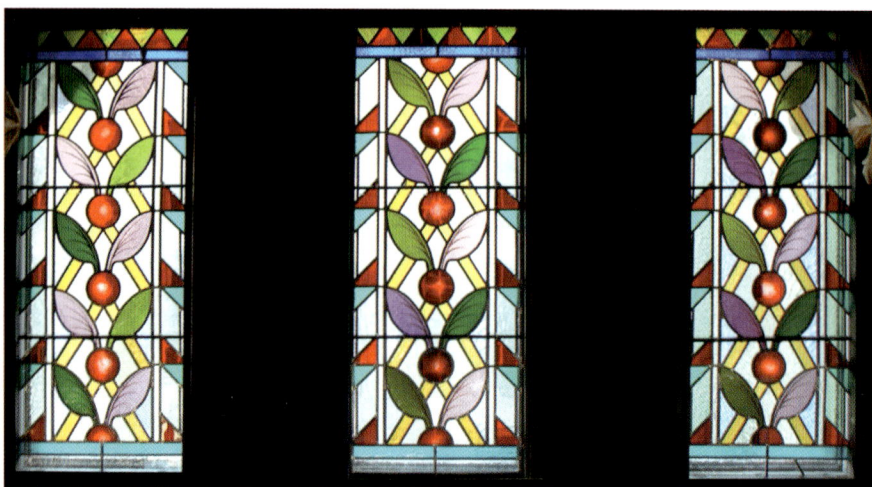

Above: John La Farge,
detail of crossing
tower window (one
of twelve), 1876,
Trinity Church,
Boston, made of
Berkshire cathedral
glass. *Roberto Rosa,
Serpentino Stained and
Leaded Glass, Needham
Heights, Massachusetts.*

Left: John La Farge,
Battle Window, 1878–
82, Memorial Hall,
Harvard University,
Cambridge,
Massachusetts. *Julie
L. Sloan.*

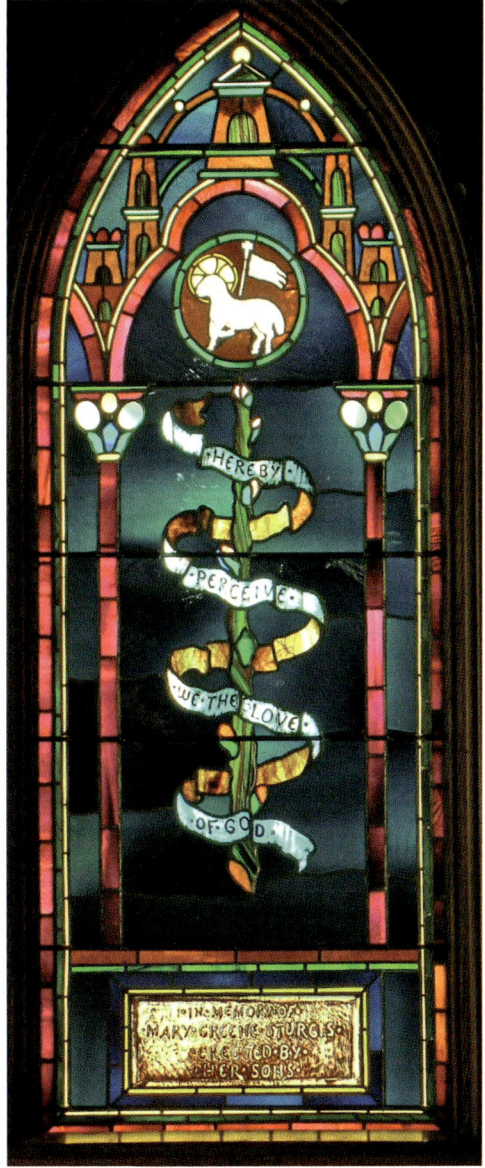

Above left: John La Farge, *Venetian Banker*, 1883–84, Christ Church, Lonsdale, Rhode Island.
Julie L. Sloan.

Above right: John La Farge, Mary Greene Sturgis Memorial, circa 1899, Emmanuel Episcopal
Church, Manchester-by-the-Sea, Massachusetts, made with teal blue antique glass in the
background, made by Berkshire Glass Works. *Julie L. Sloan.*

Tiffany Studios,
Mary Atwell
Vinton Memorial,
1883, Christ
Church, Pomfret,
Connecticut. *Julie L.
Sloan.*

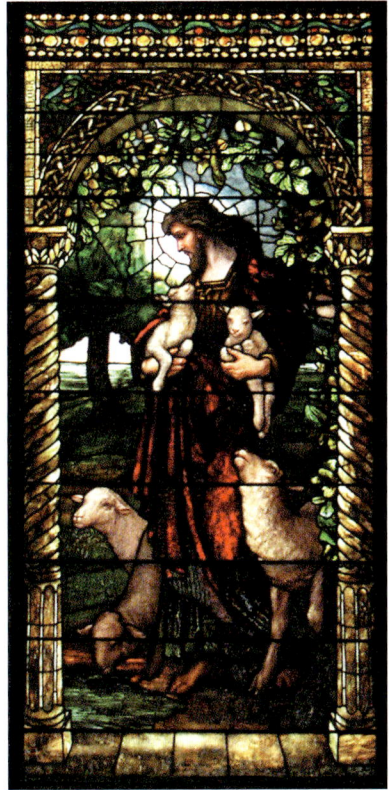

Right: Tiffany Studios, *The Good Shepherd*, circa 1889, First Presbyterian Church, Galveston, Texas. *Julie L. Sloan.*

Below: Detail, Tiffany Studios, *The Good Shepherd*, circa 1889, First Presbyterian Church, Galveston, Texas. The green-and-clear streaky glass is Berkshire glass. *Julie L. Sloan.*

Frederic Crowninshield, *The Transfiguration*, 1882, Grace Episcopal Church, New Bedford, Massachusetts. Most of the glass in this window is Berkshire glass. *Julie L. Sloan.*

Frederic Crowninshield, *Pericles and Leonardo*, 1882, Memorial Hall, Harvard University. *Julie L. Sloan.*

Detail, Frederic Crowninshield, *Pericles and Leonardo*, 1882, Memorial Hall, Harvard University. The blue glass in the figure's robe and the streaky red-and-blue glass of the building in the background were made by Berkshire Glass Works. *Julie L. Sloan.*

Prentice Treadwell, Harding Memorial, 1889, First Congregational Church, Pittsfield, Massachusetts. This window, given by William G. Harding in memory of his wife and two children, was made entirely of Berkshire glass. *Julie L. Sloan.*

Part II

STAINED GLASS

COLORED GLASS

In addition to making window glass, the Berkshire Glass factory had begun making colored glass for stained-glass windows by 1865.[211] It made both antique (blown) and cathedral (rolled) glass and was the first company in the United States to produce the latter. It is rare to be able to match a piece of glass in a stained-glass window to the factory that produced it, but thanks to the existence of middens of glass shards on the site of the Berkshire Glass Works, it is possible to identify theirs. Types and colors from the site seem to be endless but include blown and rolled teal; rolled and flashed cobalt blue; blown, rolled and crackle root-beer brown; celadon green crackle; rolled moss green; a rolled streaky red-and-clear appropriately called "beef-steak"[212]; a dichroic glass that is brick red in reflected light and teal blue in transmitted; opalescent glass the color of glacial runoff; a quadruple-flashed glass of a medium blue; cathedral tints of yellow, blue, green and pink; blown "venetian" glass with elongated concavities in purple, teal and brown; enameled glass in both clear and colored; ruby flashed, some with lines abraded through the flash; and pale green cathedral with brown streaks. As in many other glass factories, the workers also produced *whimsies*, such as canes, chains of glass, drinking glasses and witch balls, for their own use, using glass left in the pots at the end of the day. Berkshire's production of crackle glass was explained in 1871: "It is gathered the same as plain window glass, then dipped into water and afterwards burned or heated over, then blown, the result being a beautiful waving pattern all over the sheet."[213] Other types and colors of the glass are found in the houses and churches around the factory in Berkshire Village and in other local buildings. A literal palette of Berkshire glass makes up a window in the Stone Church of St. Luke's Episcopal in Lanesborough: 107 different colors are arrayed in rectangles approximately five and a half by eight inches.[214] A collection of colored glass samples was donated to the Berkshire Museum and the Berkshire Historical Society by Conrad Hines, the factory's last president.

The factory first produced cylinder antique glass, both pot-metal (colored throughout the body) and flashed (a pale tinted or clear glass with a thin layer of colored glass on one side) as early as 1865. It was awarded a bronze medal for its "improved and very fine specimens" of "colored window glass" at the Tenth Exhibition of the Massachusetts Charitable Mechanic Association in Boston.[215] By 1876, the factory was also producing rolled cathedral glass "in about 40 shades."[216] The manufacture of antique glass had been revived in England in the mid-nineteenth century. James Powell & Sons in London and James Hartley of Sunderland developed antique, or "muff," glass in the mid-1840s.[217] This was made by the same process as cylinder window glass was made. Cathedral glass was also developed in England by James Hartley, who was manufacturing it by 1850.[218] It is possible that when Harrison Page visited England in 1866 to hire glassworkers, he found one or two who had been employed by Chance Brothers, James Powell or, more likely, James Hartley and knew how to make colored glass by both blowing and rolling. By 1879, the factory had increased its palette of cathedral glass to seventy colors. The glass was made in sheets 30 by 60 inches and ⅛ inch thick, similar to double-thick window glass. It wholesaled for $0.40 per square foot (about $5.00 per sheet, or over $110.00 today), except for green, which was closer to $0.52 per square foot ($6.50 per sheet, or $145.00 today) because it was more difficult to make.[219]

In 1878, Page & Harding exhibited a stained-glass window at the Thirteenth Exhibition of the Massachusetts Charitable Mechanic Association in Boston. It is not known who designed this window or what it looked like, but its description suggests it was made at the glassworks. Shards of painted glass have been found on the site, as well as pieces that are cut in paisley shapes and other curved forms that are clearly for a stained-glass window. Awarded a silver medal, Page & Harding was described as "the only manufacturers in America" of rolled cathedral glass. "For variety and excellence of color and brilliance of surface, this glass…compares well with the products of foreign lands from which our artists and dealers have formerly obtained their supplies."[220]

The company was also making "enameled" glass, clear or colored glass with a regular pattern stenciled onto the surface in white glass paint that was then fired into the glass.[221] Enameled glass was easy to produce, providing there was a kiln in which to fire the painted glass. Other glassmakers produced it as well, and the same patterns can found around the country, so it is difficult to identify Berkshire's enameled glass in use. Several of the patterns from the glass house site are duplicated in the catalogues of other glassmakers, including an 1863 book from Chance Brothers, England, and an 1870s catalog from Carter Brothers, American Stained Glassworks in Pittsburgh. The same designs appear as late as 1923.[222] Two of the patterns were used in a hanging lamp in Chesterwood, the Stockbridge, Massachusetts home of the sculptor Daniel Chester French.[223]

Gaffield provides some of the most detailed information on Page & Harding's colored glass. In October 1877, Gaffield reported that Page & Harding was "now making brilliant blue, green & yellow rolled cathedral glass, about ⅛ in. thick," in its Lanesborough factory:

> [Page] *thinks that it is the fine sand which makes his color more brilliant than the English article. Mr. P's son* [Walter] *told me that in making colored glass of the colors named different shades were made by first putting a considerable coloring matter into colored batch to be melted in the pot, then working the pot out as nearly as usual, then filling the pot again with colorless batch & when melted, working it out again, making a glass of lighter shade than the first. A third operation in the same way will produce a third pot of still lighter shade of the same color. The first working is called Pot Metal, & the subsequent ones Tints or Tinted glass.*[224]

A few years later, he wrote:

> *Mr. Schaff told me that he had difficulty in making a uniform tint of olive green. He will get the color right in making a first melt, & then in repeating the same batch in the same pots, he will find the color more yellow than the first. He thought that the manganese might have caused the difficulty in* [?]-*joining the iron…*
>
> [Schaff] *told me of the difficulties experienced in making certain olive tints of cathedral glass, & of the ease with which he could generally produce any desired tint of blue & purple. He makes green of different tints by means of copper & chromium. He makes yellows & buffs & gold colors by different portions of sawdust. Mr. S. uses sawdust made of poplar wood. I suppose the other woods would also answer the purpose…*
>
> *I obtained a specimen of rolled cathedral glass which showed a mottled surface of two or three colors, reddish & olive green. I saw in Boston, at*

Page & Harding's shop, another similar specimen which had a kind of mottled bloody color. This latter was made of oxide of iron, copper & sawdust with white cullet upon the top of the pot. The double & triple colored pieces are the results of peculiar circumstances in the melting or [?] of gathering or casting of the glass.[225]

In 1877, a peculiar quasi-medical finding by General Augustus J. Pleasanton announced that exposure to light rays filtered through blue glass—in particular, a color called "mazarine" blue—cured everything from hair loss to measles, even reviving a canary's lost voice, and promoted literature and reading.[226] "The invalid is recommended to take an air bath in a room having the sunlight coming through this blue glass."[227] Panes of blue glass were sold framed or unframed to suspend in a south-facing window, and soon blue-glass lamp chimneys flooded the market. "A glass company on the Berkshire Hills" was reported to be "receiving orders for blue glass from all over the country."[228] This company had to be the Berkshire Glass Company, which was the only glass factory in Berkshire County that made colored glass. Within a month, the Berkshire Glass Works reported making three thousand feet of blue glass per day, while another paper reported that it made sixteen thousand feet in a week.[229] Page & Harding's Boston store became "quite a center of attraction for all seekers after the genuine article."[230] This "genuine article" was colored throughout the mass of the glass rather than flashed (or layered) on the surface, which supposedly made it superior for therapeutic purposes. Its efficacy was attested to in the letters received by the Boston store from satisfied (and apparently cured) customers. The factory's experience in making colored window and cathedral glass had "thus prepared [it] to lead all other factories in the production of the mazarine shade of blue."[231]

Thomas Gaffield maintained that the medical finding was "removed beyond the sphere of criticism and placed among the many melancholy burlesques of science and inductive investigations which have already become notorious."[232] Nonetheless, he found out that Page was indeed making blue glass at this amazing rate and that he was selling it for $0.20 to $0.25 ($4.00 to $5.00 today) per square foot wholesale (it retailed for $0.50 to $0.70 per square foot—$10.00 to $16.00 today—which is fairly expensive).[233] Trade journals and newspapers confirmed that blue glass was selling for twice the price of window glass, due to the demand, although Page & Harding also maintained that the unspecified colorant "cost five times as much as all the material for a batch of plain glass."[234] Gaffield reported that Page devoted an entire eight-pot furnace to it and used cobalt or smalt, which accounted for the great cost of the glass.[235] Page later told Gaffield that he had also made blue glass with copper oxide, "using bone lime to keep it from shading to green."[236]

The early 1880s were the strongest period for Berkshire's cathedral glass. Two furnaces were devoted to the manufacture of colored glass, one for rolled cathedral and the other for blown or cast glass, including "antique, crackled, opalescent and venetian varieties."[237] It was hailed as rich and beautiful, able to compete successfully with imported European glass. Demand was high, owing, it was said, to the increasing popularity of stained glass in residences.[238] The trade journals reported,

> *Fancy and colored glass generally known as rolled, cathedral and antique, met with a good call, and prices advanced during the closing three months [of 1883] in sympathy with other glass. The manufacture of colored glass in this country was enhanced by the starting of a large factory at Boston, but not enough is made to supply the trade. The year closes with ample stocks in the hands of jobbers and very favorable prospects for a large business next year.*[239]

By 1887, however, the factory had a strong competitor for cathedral glass. The Mississippi Glass Company of St. Louis, Missouri, crowed that it had taken the "lead of all for church and house decorative purposes. The other factory is in Massachusetts, but [it] lacks the advantages or materials and eligible location possessed by the St. Louis establishment."[240] By 1893, Mississippi Glass boasted that "the major portion of…cathedral and other rolled glass in colors, is produced by this company" and listed nine types of cathedral patterns, six of which were patented.[241]

Stained-Glass Windows
Made with Berkshire Glass

Berkshire glass found its way most commonly into windows from Boston stained-glass studios in the 1870s. When the rage for opalescent windows (those made with milky-looking glass) began in about 1880, leading to a skyrocketing demand for windows in residential buildings, New York artists, including Louis Comfort Tiffany and John La Farge, began using it as well.

W.J. McPherson Co. and Donald MacDonald

The earliest surviving window to include Berkshire glass was made in 1872. *Charity and Devotion*, installed in St. Anne's Episcopal Church in Lowell, Massachusetts, north of Boston, was designed by Donald MacDonald (1841–1916), who continued to use Berkshire glass for a number of years in windows of his own design and those he fabricated for other designers, including John La Farge (1835–1910) and Frederic Crowninshield (1845–1918).[242] MacDonald was employed by William J. McPherson and Co., a decorative art company in Boston, from 1872 until 1877, when he left to open his own studio. *Charity and Devotion* is a two-lancet window with a single figure in each lancet. Blown venetian glass from the Berkshire factory creates the blue backgrounds in the lancet heads and the red borders.[243] Red and blue venetian glass and clear ribbed glass are found in other windows made at the McPherson shop, including door lights in Harvard University's Memorial Hall (circa 1874) and decorative windows in the Parish Hall of Unity Church, North Easton, Massachusetts (1875).[244]

One of McPherson Co.'s largest stained-glass projects in which Berkshire glass was used was the glazing program for the State Capitol in Hartford, Connecticut (1878).[245] The Aesthetic-style windows illuminate the Legislative Chambers and the Hall of Flags and decorate many of the doors into offices. Laylights in the Judiciary, the main halls and the stairs complete the series.

After leaving McPherson's employ, MacDonald continued to make stained-glass windows on his own, many of them with Berkshire glass, including his 1890s set of windows in the Flint Library in Middleton, Massachusetts.

COOK, REDDING AND REDDING, BAIRD

The Boston studio of Cook, Redding (1874–83) and its successor, Redding, Baird (1883–circa 1905) created windows of Berkshire cathedral glass.[246] Cook, Redding's style was Aesthetic, and the studio's designs were mostly decorative, not figural. Redding, Baird carried on this style for a few years but also did figural designs, eventually taking up the opalescent style by the end of the 1880s. Two buildings in upstate New York are notable for their windows of Berkshire cathedral glass by these two firms.

The Waterloo Library (Waterloo, New York) was fully glazed with ornamental windows from Cook, Redding, in 1882.[247] The palette is of dusty rose, moss green, robin's egg blue and pale straw, with highlights of ruby and cobalt. The glass is predominantly Berkshire cathedral, with touches of flashed antique glass that may or may not be from Berkshire.

Beginning in 1883, the First Congregational Church of Malone, New York, purchased a large set of windows from Redding, Baird, with two signed figural windows. The palette is similar to that of the Waterloo Library.

WILLIMANTIC LINEN CO., MILL NUMBER 4

In 1880, colored glass from Berkshire was installed in Mill Number 4 of Willimantic Linen Company in Willimantic, Massachusetts.[248] Set in the transoms above the large clear-glass windows, the square, rectangular and diamond-shaped panes provided a bit of cheerful color. Now lost, such a use of this glass was probably quite common.

JOHN LA FARGE

A number of windows by John La Farge contain Berkshire glass. Donald MacDonald fabricated La Farge's windows until about 1880, and this relationship is probably how La Farge became familiar with the glass. La Farge continued to employ it in his windows for years after he stopped working with MacDonald. The earliest surviving windows made by MacDonald for La Farge were for the latter's circa 1875 commission for the Walter Hunnewell house in Wellesley, Massachusetts.[249] A group of two small panels of circular quarries and one of square quarries, they were originally installed in an inglenook beneath the main stairs.[250] Their form is based on late medieval northern European residential glass. They were decorated with tiny paintings copied not from medieval European sources, as might be expected, but from the *Manga*, a group of fifteen notebooks of the Japanese artist Katsushika Hokusai (1760–1849). The glass is rolled Berkshire glass in pale tints of blue, green, yellow and pink.

La Farge designed another early group in 1876 for the crossing tower of Trinity Episcopal Church in Boston by architect H.H. Richardson. These windows are composed of Berkshire cathedral glass in strong tones of red, orange, green and purple. The design is a simple one. In the lower paired openings, a field of conventionalized papyrus with a delicate tracing of veins is set within a border of chevrons. Above, three vertical transoms contain abstracted flowers and leaves. These windows were fabricated by Samuel West, a Boston stained-glass artist.[251]

It can be difficult to recognize individual pieces of glass in La Farge's windows after 1878 because they are assembled in numerous layers of glass that obscure the characteristics of specific pieces. One of the most easily identifiable colors, however, is a dark mottled red cathedral, so deeply colored that in all but the brightest direct sunlight it appears almost black. La Farge used this glass for the back of the main figure in the right lancet of the *Battle Window* (1877–82) in Memorial Hall at Harvard University; in one of the decorative aisle windows in the Newport Congregational Church (1880–81); and as the robes of the *Venetian Banker* in the chancel of Christ Episcopal Church in Lonsdale, Rhode Island (1883–84).

Another Berkshire color that La Farge used a great deal was a blown glass of deep teal made with copper. It forms the background of the memorial windows to Mary Greene Sturgis and Abby Sears McColloh in Emmanuel Episcopal Church in Manchester-by-the-Sea, Massachusetts, made in 1899. Another easily recognizable Berkshire glass is a clear cathedral with streaks of dark mossy green and brown, giving the glass the appearance of moss agate. It forms the grassy hills in La Farge's 1886 chancel windows in Trinity Episcopal Church, Buffalo, New York.

TIFFANY STUDIOS

Windows by Tiffany Studios made through the end of the 1880s, if not later, also contain a great deal of Berkshire glass.[252] Some of the earliest are the windows of St. Stephen's Episcopal Church in Lynn, Massachusetts, created in 1881. Three large windows are complemented by many small, double-hung windows around the church. In at least one of these can be seen a streaky cream-and-brown blown glass from Berkshire.

A remarkable set of six windows created in 1881–83 for Christ Episcopal Church, Pomfret, Connecticut, by Tiffany incorporates a plethora of Berkshire glass.[253] All donated by a single family in memory of their forebears, the windows exhibit Louis Comfort Tiffany's desire to create windows without using glass paint.[254] To that end, vividly figured glass was artistically useful. Deep mottled red, heavily rippled butterscotch and dark teal blue Berkshire glasses were used in the rose. In the Mary Atwell Vinton Memorial, brown-streaky- and green-streaky-on-clear glass forms the ground under the woman's feet. Behind her head, a tumultuous sky is rendered in blue-and-red dichroic glass.

The "moss agate" glass that La Farge used in his 1886 windows in Trinity Episcopal Church, Buffalo, also forms the pool at the feet of Jesus in the *Good Shepherd* in the First Presbyterian Church in Galveston, Texas (circa 1889).[255] Tiffany also utilized the deep mottled red in the robes of Jesus.

FREDERIC CROWNINSHIELD

Frederic Crowninshield (1845–1918) designed stained-glass windows from 1879 until about 1914. He lived in Boston in the 1870s and first half of the 1880s, and his windows from this period were fabricated by Donald MacDonald. Most, if not all, from this period incorporate Berkshire glass. It was probably MacDonald who introduced Crowninshield to Berkshire glass, even though Crowninshield knew the Berkshires as early as 1880, buying a summer home in Stockbridge, Massachusetts, in 1893. He was said to have visited the glassworks, but no date of such a visit was given.[256] In 1885, Crowninshield moved to New York, where he fabricated his windows himself, and Berkshire glass was no longer commonly found in them.

Crowninshield's first two windows were for the chapel of the First Church (Unitarian) in Boston, created in 1879 and destroyed by fire in 1968. All that remains of these windows is a written description and a sketch of one of them. One of these windows, *St. Christopher Fording a Stream with the Infant*

Christ on His Shoulders, was said to have been made of cathedral glass.[257] Given the early date and MacDonald as manufacturer, it seems likely that it was made with Berkshire glass.

The next project Crowninshield received was for fifteen windows in Grace Episcopal Church in New Bedford, Massachusetts. The windows were completed in 1881 and 1882. The largest, *The Transfiguration*, is composed almost exclusively of Berkshire glass. The turbulent sky is of streaky white and blue glass, the deeper tints at the top fading to a lighter tone near the horizon. The figures of the apostles cowering in the foreground are depicted with a deep tan glass, and they are dressed in streaky olive green or brown. The robes worn by the prophets are of clear glass streaked with brown and teal. The halos are a stunning deep orangey-red. The background of the surrounds is a streaky yellow, amber and brown glass. While not the most vibrant palette, the tumultuous swirls in the glass provide the scene with drama.

One of Berkshire's most interesting colors is a dichroic glass, turquoise blue in transmitted light and deep cordovan red in reflected light. Crowninshield's first two windows for Harvard University's Memorial Hall make good use of this glass as clothing for the figures. In the earlier of the windows, *Pericles and Leonardo* (1882), this glass appears in Pericles's breastplate and in the skirt of Leonardo's tunic. In the second window, *Sophocles and Shakespeare* (1883), it forms Sophocles's mantle and Shakespeare's cloak. An extraordinary streaked glass, it is also full of bubbles called "seeds."

E. PRENTICE TREADWELL

In 1889, William G. Harding commissioned a stained-glass window in memory of his wife and two children—who had died, it will be remembered, in 1874—for First Church (Congregational) in Pittsfield, designed by Ezra Prentice Treadwell (1848–1903) and made of Berkshire glass.[258] Treadwell was a designer working in many media but mostly as an interior decorator.[259] Based in Boston, he had decorated the chapel of First Church in 1882.[260]

The Harding memorial window comprises two lancets, each containing a woman with two small children. In the left lancet, the woman faces forward, looking off to her left, with her right hand on the taller child's head and a book in her left hand. The children, representing Hope and Malcolm Harding—who were three and five years old, respectively, at the time of their deaths—have their backs to the viewer, looking at the woman, their mother. They stand on green-brown earth, signifying their living presence. The woman's face is painted in fine detail and appears to be a portrait of Nannie Harding. All are dressed in vaguely classical garb. In the right lancet, the woman holds both children, whose heads rest on her shoulders, and has turned her back to the viewer in an allegory of death. There is no earth below her feet. Now she stands on pinkish clouds.

The glass used in the window is unquestionably Berkshire glass, matching shards dug at the factory site. In both lancets, Nannie Harding's dress is made of a streaky red on pale yellow. In the left lancet, there is more yellow in the glass, while on the right, deep red predominates. The flesh tones are a surprisingly dark tan. The grass in the right lancet is a streaky moss-green-and-brown in clear. The same glass creates a wide border in the dedication panels. The sky behind the figures blends from pale yellow-pink into a brilliant turquoise streaked with blue and gold. The glass company's famous mazarine blue forms the drape over the living Nannie's arm and the background to the dedications at the bottom.

Treadwell created another stained-glass window around this time, for the State Normal College in Albany, New York. Commissioned in 1886 and destroyed by fire in 1906, this window depicted life-size allegorical figures representing "the development of mankind and of the arts and sciences as the result of education." Treadwell wrote that he would make the window "in mosaic of American colored glass, all the shading of color in figures, faces and draperies being carefully painted on and burned into the glass, while the outline drawing will be entirely of lead." Said to be the largest stained-glass window in the country at $32\frac{1}{2}$ feet tall by $14\frac{1}{2}$ feet wide, the window was not completed until 1892, due to protracted fundraising.[261] The notice of its completion cited the earlier description, adding, "It will be observed that the window is to be in a 'mosaic of American colored glass,' thus incidentally showing the progress in artistic industries in the United States; this peculiar work, now coming into very general use, could hardly have been obtained ten years previously."[262]

This description suggests that the window was made of antique and/ or cathedral glass, rather than opalescent glass (such as that used by Louis Comfort Tiffany, which was coming into vogue and availability at this time), because the claims for opalescent glass held that it did not require painting.[263] The Harding memorial window is painted in this way and contains no opalescent glass.

* * *

The last building to be glazed with Berkshire glass was probably the Catholic Church of the North American Martyrs in Berkshire Village, built in 1935 (now closed). The dormer and sacristy windows were made with Berkshire glass that had been stored by a resident of the village.[264]

Although the glass is no longer commercially available, windows are still being made and restored with pieces salvaged and excavated from around the village.

Appendix I

LIST OF BUILDINGS WITH BERKSHIRE GLASS

1868, Kent Building, New Hampshire Asylum for the Insane, Concord, New Hampshire (window glass)[265]

circa 1870s, St. James Episcopal Church, Great Barrington, Massachusetts

1872–79, roofs of the train sheds of the Boston & Lowell and Eastern Railroads in Boston[266]

1872, Donald MacDonald, *Charity and Devotion*, St. Ann's Episcopal Church, Lowell, Massachusetts

1875, W.J. McPherson Co., parish house windows, Unity Church, North Easton, Massachusetts

circa 1875, John La Farge, *Manga*, Walter Hunnewell house, Wellesley, Massachusetts

1876–77, John La Farge and Samuel West, crossing tower windows, Trinity Episcopal Church, Boston, Massachusetts

1876–78, New York County (Tweed) Courthouse, New York, New York, designed by John Kellum (1809–1871) and Leopold Eidlitz (1823–1908), designed 1861–81[267]

1877–82, John La Farge, *Battle Window*, Memorial Hall, Harvard University

1878, W.J. McPherson Co., State Capitol, Hartford, Connecticut

1879, Frederic Crowninshield, *St. Christopher Fording a Stream with the Infant Christ on His Shoulders*, First Church (Unitarian), Boston, Massachusetts

1880, Willimantic Linen Co., Mill Number 4

1880–81, John La Farge, Congregational Church, Newport, Rhode Island

circa 1880–98, St. Luke's Episcopal Church, Lanesborough, Massachusetts (both summer and winter churches)[268]

1881–83, Tiffany Studios, Vinton family memorial windows, Christ Episcopal Church, Pomfret, Connecticut

1881, Tiffany Studios, St. Stephen's Episcopal Church, Lynn, Massachusetts

1882, Frederic Crowninshield, *Pericles and Leonardo*, Memorial Hall, Harvard University

1882, Frederic Crowninshield, *Musical Angels* (destroyed), *Floral Windows*, *Four Evangelists* and *The Transfiguration*, Grace Episcopal Church, New Bedford, Massachusetts

1882, Cook, Redding, Waterloo Library, Waterloo, New York

1883, Frederic Crowninshield, *Sophocles and Shakespeare*, Memorial Hall, Harvard University

1883–84, Redding, Baird, *Moses* and other windows, First Congregational Church, Malone, New York

1883–84, John La Farge, *Venetian Banker*, Christ Episcopal Church, Lonsdale, Rhode Island

1884, Tiffany Studios, lunettes, Catholic Church of the Sacred Heart, New York, New York[269]

1886, John La Farge, chancel windows, Trinity Episcopal Church, Buffalo, New York

1886, E. Prentice Treadwell, alumni window, State Normal College, Albany, New York (destroyed)

1886, Samuel West, City Hall, Holyoke, Massachusetts

1887, State Library, Austin, Texas (plate glass)

1888, Metropolitan Museum of Art, New York, New York (plate glass)

1889, E. Prentice Treadwell, Harding memorial window, First Congregational Church, Pittsfield, Massachusetts

circa 1889, Tiffany Studios, *Good Shepherd*, First Presbyterian Church, Galveston, Texas[270]

1890s, Donald MacDonald, Flint Library, Middleton, Massachusetts

1899, John La Farge, Sturgis and McColloh memorial windows, Emmanuel Church, Manchester-by-the-Sea, Massachusetts

1935, Catholic Church of the North American Martyrs, Berkshire Village, Lanesborough, Massachusetts

Appendix II
RESIDENTS OF
BERKSHIRE VILLAGE

The following is a list of residents recorded by the U.S. Census in 1850, 1860, 1870, 1880 and 1900, during the period when the Berkshire Glass Works was in operation. Names in bold are heads of households.

Spellings in the census vary from year to year and should not be relied upon. For example, Beckley will be spelled Beckly and Callahan as Calohan. Mary is often listed as Maria, and Alonso alternates with Alonzo.

Other employees of the glassworks may not have lived in Berkshire Village. The censuses of Pittsfield, Lanesborough, New Ashford and Cheshire were not surveyed for glassworkers.

FIRST NAME	LAST NAME	RACE	OCCUPATION	ORIGIN	CENSUS YEAR
Clara	Acker	White	Domestic	Massachusetts	1870
Owen	**Agan**	White	Farm laborer	Ireland	1860
Ellen	**Agen**	White	Domestic	Ireland	1860
Jeremiah	**Ahaire**	White	Farm laborer	Ireland	1860
Margaret	Ahaire	White	Wife	Ireland	1860
Elizabeth	Ahaire	White	Child	New York	1860
Mary A.	Ahaire	White	Child	New York	1860
Julia	Ahaire	White	Child	Massachusetts	1860
Aglan	**Aimable**	White	Glass factory laborer	Belgium	1880
Ernest	Aimable	White	Child		1880
Henry	**Aimable**	White	Glass factory laborer	Belgium	1880

First Name	Last Name	Race	Occupation	Origin	Census Year
Lasnie	**Aimable**	White	Glass factory laborer	Belgium	1880
John	**Ambler**	White	Clergy, Baptist	New York	1860
Abagail F.	Ambler	White	Wife	New York	1860
Julius	**Amede**	White	Farm laborer	Massachusetts	1900
Lousa	**Andries**	White	Glass factory laborer	Belgium	1880
Henry	**Antoine**	White	Glass blower	Germany-Bohemia	1880
Louise	Antoine	White	Wife		1880
Margaret	Archer	White	Child	Massachusetts	1860
Harriet	**Arkin**	White	Child	Canada	1860
William	**Ashman**	White	Glass blower	Canada	1860
James	Bagg	White	Child		1850
James	Baker	White	Child		1850
John	Baker	White	Child		1850
William	**Baker**	White	Glass factory laborer	New York	1870
Mary	Baker	White	Wife		1870
Mary	Baker	White	Child		1870
William	Baker	White	Child		1870
Fred	Baker	White	Child		1870
Sarah	Baker	White	Child		1870
Daniel	Baker	White	Child		1870
William	**Baker**	White	Day laborer	Vermont	1900
Agnes	Baker	White	Wife		1900
William	**Baker**	White	Day laborer	Massachusetts	1900
Agnes	Baker	White	Wife		1900
Freddie	Baker	White	Child		1900
Mary	Baker	White	Child		1900
David	Baker	White	Child		1900
Walter	Baker	White	Child		1900
Christy	Baker	White	Child		1900
Merrick	**Barnard**	White	Laborer	Massachusetts	1850
Elizabeth	Barnard	White	Wife		1850
Andre	**Barnard**	White	Glass flattener	Massachusetts	1880
Lena	Barnard	White	Wife		1880
Margie	Barnard	White	Child		1880

Residents of Berkshire Village

First Name	Last Name	Race	Occupation	Origin	Census Year
James	Barnard	White	Child		1880
(?)	**Barney**	White	Glass factory laborer	Massachusetts	1880
Josephia	Barney	White	Wife		1880
Nancy	**Barrie**	White	Domestic	Ireland	1860
Calvin	**Bartlett**	White	Carpenter	Massachusetts	1850
Stephen	**Bartlett**	White	Laborer	Massachusetts	1850
Lydia	Bartlett	White	Wife		1850
Julia	Bartlett	White	Child		1850
Walter	Bartlett	White	Child		1850
Harry	**Bassett**	White	Glass factory laborer	Massachusetts	1880
[wife]	Bassett	White	Wife		1880
Vicky (?)	Bassett	White	Child		1880
Mary	Bassett	White	Child		1880
Ennette	Bassett	White	Child		1880
Thomas	**Battles**	White	Glass gatherer	England	1870
Eliza	Battles	White	Wife		1870
Sarah	Battles	White	Child		1870
Elizabeth	Battles	White	Child		1870
Mary	Battles	White	Child		1870
William	Battles	White	Child		1870
Peter	Battles	White	Child		1870
Joseph	**Beckley**	White	Glass cutter	New Jersey	1860
Caroline	Beckley	White	Wife	New York	1860
Henry H.	Beckley	White	Child	New York	1860
Edwin	Beckley	White	Child	New York	1860
Ida	Beckley	White	Child	Massachusetts	1860
Joseph	**Beckly**	White	Glass cutter	New Jersey	1870
Caroline	Beckly	White	Wife		1870
Hasbrouck	Beckly	White	Child		1870
Edmund	Beckly	White	Child		1870
Ida	Beckly	White	Child		1870
Clarina	Beckly	White	Child		1870
Mary	**Bennett**	White	Teacher (public school)	Massachusetts	1900
Charles	**Best**	Black	Laborer	Massachusetts	1850
Clarrisa	Best	Black	Wife		1850

First Name	Last Name	Race	Occupation	Origin	Census Year
Sarah	Best	Black	Child		1850
Charles	Best	Black	Child		1850
Charles	**Best**	White	Glass factory laborer	Bavaria	1860
Mary	Best	White	Wife	Bavaria	1860
Louisa	Best	White	Child	Bavaria	1860
Sophia	Best	White	Child	New York	1860
Hellen	Best	White	Child	Massachusetts	1860
Michael	**Best**	White	Glass factory laborer	Bavaria	1860
Margaret	Best	White	Wife	Bavaria	1860
Catharine	Best	White	Child	Bavaria	1860
Margaret	Best	White	Child	Bavaria	1860
Louis	Best	White	Child	New York	1860
William	Best	White	Child	New York	1860
Louisa	Best	White	Child	New York	1860
Charles	Best	White	Child	New York	1860
John	Best	White	Child	New York	1860
Charles	**Best**	White	Glass flattener	Germany-Wirtemberg	1870
Maria	Best	White	Wife		1870
Sophia	Best	White	Child		1870
Charles	Best	White	Child		1870
Rosina	Best	White	Child		1870
Julia	Best	White	Child		1870
Frank	Best	White	Child		1870
Joseph	Best	White	Child		1870
Louise	Best	White	Domestic	Germany-Wirtemberg	1870
Charles	**Best**	White	Glass works	England	1880
Mary	Best	White	Wife		1880
Sophiah	Best	White	Child		1880
Charles	**Best**	White	Glass works	Massachusetts	1880
Julia	Best	White	Wife		1880
Frank	Best	White	Child		1880
Joseph	Best	White	Child		1880
Charles	**Best**	White	Glass works	England	1880
Mary	Best	White	Wife		1880

Residents of Berkshire Village

First Name	Last Name	Race	Occupation	Origin	Census Year
Sophiah	Best	White	Child		1880
Patrick	**Best**	White	Farmer	Ireland	1880
Mrs.	Best	White	Wife		1880
Agey	Best	White	Child		1880
Owen	Best	White	Child		1880
Charles	**Best**	White	Child		1900
Mary	Best	White	Child		1900
Frank	**Best**	White	Glass gatherer	Massachusetts	1900
Mary	Best	White	Wife		1900
Lena	Best	White	Child		1900
Katherine	Best	White	Child		1900
Raymond	Best	White	Child		1900
Joseph	**Best**	White	Glass gatherer	Massachusetts	1900
Mary	Best	White	Wife		1900
Mary	Best	White	Child		1900
Anna	Best	White	Child		1900
Frances	Best	White	Child		1900
Jabez	**Blackmer**	White	Day laborer	Scotland	1900
Lizzie	Blackmer	White	Wife		1900
Samuel	**Borneshall**	White	Glass gatherer	England	1870
Elizabeth	Borneshall	White	Wife		1870
Elizabeth	Borneshall	White	Child		1870
Fannie	Borneshall	White	Child		1870
John	Borneshall	White	Child		1870
Samuel	Borneshall	White	Child		1870
Joseph	**Bouneau**	White	Glass gatherer	Canada	1870
Sarah	Bouneau	White	Wife		1870
John	Bouneau	White	Child		1870
Mary	Bouneau	White	Child		1870
Joseph	Bouneau	White	Child		1870
Rosa	Bouneau	White	Child		1870
Alfred	Bouneau	White	Child		1870
John	**Brannery**	White	Teamster	Massachusetts	1880
Hatty	Brannery	White	Wife		1880
James	Brannery	White	Child		1880
Mary	Brannery	White	Child		1880
Thomas	**Brannery**	White	Farmer	Ireland	1880

First Name	Last Name	Race	Occupation	Origin	Census Year
Bridget	Brannery	White	Wife		1880
Christopher	Brannery	White	Child		1880
Thomas [Jr.]	**Brannery**	White	Milk peddler	Massachusetts	1880
William	**Brannery**	White	Teacher	Massachusetts	1880
Christopher	**Brennan**	White	Laborer	Ireland	1850
James	**Briggs**	White	Farmer	Massachusetts	1850
Abigail	Briggs	White	Wife		1850
Theresa	Briggs	White	Child		1850
James	**Briggs**	White	Farmer	Massachusetts	1850
Eliza	Briggs	White	Wife		1850
Lewis	Briggs	White	Child		1850
Fordyce	Briggs	White	Child		1850
F.W.	**Briggs**	White	Farmer	Massachusetts	1880
Mary	Briggs	White	Wife		1880
Henry	**Briggs**	White	Glass factory laborer	Massachusetts	1880
Latimore	**Briggs**	White	Farmer	Massachusetts	1880
Clarassy	Briggs	White	Wife		1880
Maria	Briggs	White	Child		1880
Henry	Briggs	White	Child		1880
Patrick	**Brough**	White	Glass factory laborer	Ireland	1870
Bridget	Brough	White	Wife		1870
Mary	Brough	White	Child		1870
Patrick	Brough	White	Child		1870
John	**Broughton**	White	Glass blower	England	1870
William	Brown	Black	Child		1850
Charles	**Brown**	White	Glass blower	Massachusetts	1860
B (?)	**Brown**	White	Glass factory laborer	Massachusetts	1880
Sophronia	Brown	White	Wife		1880
Charles	**Brown**	White	Glass factory laborer	Massachusetts	1880
Asabel	**Buck**	White	Farmer	Massachusetts	1850
Sophia	Buck	White	Wife		1850
George	Buck	White	Child		1850
Harriet	Buck	White	Child		1850
Truman	Buck	White	Child		1850

Residents of Berkshire Village

First Name	Last Name	Race	Occupation	Origin	Census Year
Laura	Buck	White	Child		1850
Ralph	**Buck**	White	Farmer	New York	1850
Laura	Buck	White	Wife	Connecticut	1850
Susan	Buck	White	Child		1850
Harriet	Buck	White	Child		1850
Bushrod	Buck	White	Child		1850
Perkins	Buck	White	Child		1850
Mary	Buel	White	Child		1870
Eva	Buel	White	Child		1870
Harry	Buel	White	Child		1870
Charles	Buel	White	Child		1870
Edward	**Burk**	White	Glass factory laborer	Ireland	1880
Mary	Burk	White	Wife		1880
Edward	**Burk**	White	Glass factory laborer	Massachusetts	1880
John	**Burk**	White	Teamster	Massachusetts	1880
Martin	Burk	White	Wife		1880
Francis	Burk	White	Child		1880
Pat	Burk	White	Child		1880
Ellen	Burk	White	Child		1880
Thomas	**Burk**	White	Glass factory laborer	Massachusetts	1880
Charles	**Burke**	Mulatto	Farmer	New Jersey	1870
Mary	Burke	Mulatto	Wife		1870
Charles	Burke	Mulatto	Child		1870
Mary	**Burke**	White	Wife		1900
Francis	**Burn**	White	Glass factory laborer	France	1860
William	**Burnoro (?)**	White	Glass factory laborer	Belgium	1880
Elizabeth	Burnoro (?)	White	Wife		1880
Maud	Burnoro (?)	White	Child		1880
Charles	**Burt**	Black	Glass factory laborer	New Jersey	1880
Adeline	Burt	Black	Wife		1880
Charles	Burt	Black	Child		1880
Frank	Burt	Black	Child		1880
Maria	Burt	Black	Child		1880

First Name	Last Name	Race	Occupation	Origin	Census Year
Andre	**Cailier**	White	Glass factory laborer	Belgium	1880
Timothy	**Callahan**	White	Glass factory laborer	Ireland	1870
Bridget	Callahan	White	Wife		1870
Ellen	Callahan	White	Child		1870
John	Callahan	White	Child		1870
Mary	Callahan	White	Child		1870
Philip	Callahan	White	Child		1870
Margaret	Callahan	White	Child		1870
Anna	Callahan	White	Child		1870
Alexander	**Callahan**	White	Day laborer	Massachusetts	1900
James	Callahan	White	Child		1900
Dora	Callahan	White	Child		1900
Michael	**Callahan**	White	Day laborer	Ireland	1900
Mary	Callahan	White	Wife		1900
Michael	**Callaton**	White	Glass factory laborer	Massachusetts	1880
Mary	Callaton	White	Wife		1880
Katie	Callaton	White	Child		1880
Thomas	Callaton	White	Child		1880
Michael	**Calohan**	White	Glass factory laborer	Ireland	1880
Mary	Calohan	White	Wife		1880
John	Calohan	White	Child		1880
Nella	Calohan	White	Child		1880
Desgurnes	**Capole**	White	Brass maker	Belgium	1880
Emilla	Capole	White	Wife		1880
Molina	Capole	White	Child		1880
Feranelle	Capole	White	Child		1880
Napoleon	Capole	White	Child		1880
Bridget	Carley	White	Child		1880
Patrick	**Carlin**	White	Glass factory laborer	Ireland	1880
Anna	Carlin	White	Wife		1880
Thomas	Carlin	White	Child		1880
John	Carlin	White	Child		1880
Pat	Carlin	White	Child		1880

Residents of Berkshire Village

First Name	Last Name	Race	Occupation	Origin	Census Year
James	Carlin	White	Child		1880
Peter	Carlin	White	Child		1880
William	Carlin	White	Child		1880
John	**Carlin**	White	Glass gatherer	Massachusetts	1900
Patrick	**Carlin**	White	Farmer	Ireland	1900
Richard	**Carlin**	White	Farm laborer	Massachusetts	1900
Robert	Carlin	White	Child		1900
William	**Carlin**	White	Farm laborer	Massachusetts	1900
Patrick	**Carlon**	White	Glass factory laborer	Ireland	1870
Ann	Carlon	White	Wife		1870
Thomas	Carlon	White	Child		1870
John	**Carney**	White	Glass factory laborer	Ireland	1870
Elizabeth	Carney	White	Wife		1870
Bridget	Carney	White	Child		1870
Jerry	Carney	White	Child		1900
Henry	**Carny**	White	Glass factory laborer	Ireland	1880
John	**Carny**	White	Glass worker	Ireland	1880
Elizabeth	Carny	White	Wife		1880
Ann	Carny	White	Child		1880
Sarah	Carny	White	Child		1880
John	Carny	White	Child		1880
Calvin	**Carpenter**	White	Farm laborer	Massachusetts	1860
Eliza	Carpenter	White	Wife	Massachusetts	1860
Sanford A.	Carpenter	White	Child	Massachusetts	1860
William	**Carpenter**	White	Engineer	Massachusetts	1880
Mary	Carpenter	White	Wife		1880
John	**Carro**	White	Glass factory laborer	Canada	1860
Mary	Carro	White	Wife	Canada	1860
John	Carro	White	Glass factory laborer	Canada	1860
Francis	Carro	White	Glass factory laborer	Canada	1860
David	Carro	White	Glass factory laborer	Canada	1860
Joseph	Carro	White	Child	Canada	1860

First Name	Last Name	Race	Occupation	Origin	Census Year
Rosa	Carro	White	Child	Canada	1860
Arianna	Carro	White	Child	Massachusetts	1860
John [Jr.]	**Carro**	White	Glass factory laborer	Canada	1860
Mary	**Carrol**	White	Housekeeper		1880
Morris	**Carroll**	White	Teamster	Ireland	1860
Ellen	Carroll	White	Wife	Ireland	1860
Catharine	Carroll	White	Child	Connecticut	1860
Hannora	Carroll	White	Child	Massachusetts	1860
Ellen	Carroll	White	Child	Massachusetts	1860
John	Carroll	White	Child	Massachusetts	1860
Francis	**Carroll**	White	Glass factory laborer	Canada	1870
Rose	Carroll	White	Wife		1870
Mary	Carroll	White	Child		1870
Rosa	Carroll	White	Child		1870
Joseph	Carroll	White	Child		1870
Napoleon	Carroll	White	Child		1870
John	**Carroll**	White	Glass factory laborer	Canada	1870
Mary	Carroll	White	Wife		1870
John	**Carroll**	White	Glass factory laborer	Canada	1870
David	Carroll	White	Glass factory laborer	Canada	1870
Christen	Carroll	White	Wife		1870
Mary	Carroll	White	Child		1870
Holly	Carroll	White	Child		1870
Joseph	Carroll	White	Child		1870
David	**Carroll**	White	Glass factory laborer	Canada	1880
Christiny	Carroll	White	Wife		1880
Mary	Carroll	White	Child		1880
Joseph	Carroll	White	Child		1880
Edward	Carroll	White	Child		1880
Liva [?]	Carroll	White	Child		1880
Joseph	**Carrow**	White	Glass gatherer	Canada	1870
Sophia	Carrow	White	Wife		1870
Joseph	Carrow	White	Child		1870

Residents of Berkshire Village

First Name	Last Name	Race	Occupation	Origin	Census Year
Mary	Carrow	White	Child		1870
Joseph	**Carrow**	White	Glass blower	Massachusetts	1880
Sofia	Carrow	White	Wife		1880
Joseph	Carrow	White	Child		1880
Minnie	Carrow	White	Child		1880
Charles	Carrow	White	Child		1880
Walter	Carrow	White	Child		1880
Frank	Carrow	White	Child		1880
Emmie	Carrow	White	Child		1880
John	Carrow	White	Child		1880
Charles	**Carrow**	White	Glass blower	Massachusetts	1900
Charles	**Carrow**	White	Glass gatherer	Massachusetts	1900
David	**Carrow**	White	Day laborer	Canada	1900
Edward	**Carrow**	White	Day laborer	Massachusetts	1900
Edward	**Carrow**	White	Glass gatherer	Massachusetts	1900
Mary	Carrow	White	Wife		1900
Emmie	Carrow	White	Child		1900
Anna	Carrow	White	Child		1900
James	Carrow	White	Child		1900
Eli	**Carrow**	White	Day laborer	Massachusetts	1900
Annie	Carrow	White	Wife		1900
Frank	**Carrow**	White	Glass snapper	Massachusetts	1900
Fred	**Carrow**	White	Glass snapper	Massachusetts	1900
John	**Carrow**	White	Day laborer	Massachusetts	1900
Joseph	**Carrow**	White	Glass blower	Canada	1900
Carrie	Carrow	White	Wife		1900
Edward	Carrow	White	Child		1900
Albert	Carrow	White	Child		1900
Lillian	Carrow	White	Child		1900
George	Carrow	White	Child		1900
Joseph	**Carrow**	White	Glass gatherer	Massachusetts	1900
William	**Carrow**	White	Glass gatherer	Massachusetts	1900
Andrew (?)	**Cartein**	White	Glass factory laborer	Massachusetts	1880
Martin	**Carty**	White	Glass factory laborer	Ireland	1860
Maria	Carty	White	Wife	Ireland	1860
Michael	Carty	White	Child	Massachusetts	1860

First Name	Last Name	Race	Occupation	Origin	Census Year
Maria	Carty	White	Child	Massachusetts	1860
Mary	**Carty**	White	Domestic	Ireland	1860
Martin	**Carty**	White	Glass factory laborer	Ireland	1880
Maria	Carty	White	Wife		1880
Mitre (?)	Carty	White	Child		1880
Martin	Carty	White	Child		1880
Pat	Carty	White	Child		1880
Maria	Carty	White	Child		1880
Mary	Carty	White	Child		1880
Margaret	Carty	White	Child		1880
Lawrence	Carty	White	Child		1880
George	**Cassin**	White	Laborer	Canada	1850
Eugene	**Cavelin**	White	Glass blower	Belgium	1880
Philip	**Chapins**	White	Glass factory laborer	France	1880
Justine	Chapins	White	Wife		1880
Mary	Chapins	White	Child		1880
Harvey	**Chase**	White	Farmer	Massachusetts	1850
Sarah	Chase	White	Wife		1850
Emily	Chase	White	Child		1850
William	Chase	White	Child		1850
Lawrence	Chase	White	Child		1850
Achsah	Chase	White	Child		1850
Harvey	**Chase**	White	Farmer	Massachusetts	1860
Sarah	Chase	White	Wife	Massachusetts	1860
Emily	Chase	White	Child	Massachusetts	1860
Lawrence	Chase	White	Farmer	Massachusetts	1860
Maria A.	Chase	White	Child	Massachusetts	1860
Lawrence	**Chase**	White	Farm laborer	Massachusetts	1860
Harvey	**Chase**	White	Farmer	Massachusetts	1870
Sarah	Chase	White	Wife		1870
Alonzo	Chase	White	Child		1870
Maria	Chase	White	Child		1870
Charlotte	**Chase**	White	Wife		1880
Lily	Chase	White	Child		1880
Harvey	**Chase**	White	Farmer	Massachusetts	1880
Sarah	Chase	White	Wife		1880

Residents of Berkshire Village

First Name	Last Name	Race	Occupation	Origin	Census Year
Lawrence	**Chase**	White	Farmer	Massachusetts	1880
Angelia	Chase	White	Wife		1880
William	Chase	White	Child		1880
Bido	Chase	White	Child		1880
Lawrence	**Chase**	White	Farmer	Massachusetts	1900
Amelia	Chase	White	Wife		1900
Lawrence	Chase	White	Child		1900
John	**Clark**	White	Laborer	Massachusetts	1850
Pamelia	Clark	White	Wife		1850
Stephen	**Clark**	White	Teamster	Massachusetts	1860
John	**Clark**	White	Pauper		1880
Mary	**Coleman**	White	Child		1870
William	Coleman	White	Railroad laborer		1870
Mary	Coleman	White	Child		1880
Truman	**Coman**	White	Farmer	Massachusetts	1850
Sarah M.	Coman	White	Wife		1850
Sarah A.	Coman	White	Child		1850
Richard	Coman	White	Child		1850
James	**Conaslary**	White	Glass works	Belgium	1880
Kate	Conaslary	White	Wife		1880
William	Conaslary	White	Child		1880
Micha	Conaslary	White	Child		1880
John	Conaslary	White	Child		1880
Mary	Conaslary	White	Child		1880
Nellie	Conaslary	White	Child		1880
Julia	Conaslary	White	Child		1880
Maggie	Conaslary	White	Child		1880
James	**Conelly**	White	Carpenter	Ireland	1860
Harris	**Conlan**	White	Farmer	Massachusetts	1870
James	**Connelly**	White	Carpenter	Ireland	1870
Kate	Connelly	White	Wife		1870
Anastasia	Connelly	White	Child		1870
William	Connelly	White	Child		1870
Michael	Connelly	White	Child		1870
James	**Connelly**	White	Carpenter	Ireland	1900
Kate	Connelly	White	Wife		1900

First Name	Last Name	Race	Occupation	Origin	Census Year
John	**Connelly**	White	Glass cutter	Massachusetts	1900
Ellen	Connelly	White	Wife		1900
Julia	Connelly	White	Child		1900
Margaret	Connelly	White	Child		1900
James	Connelly	White	Child		1900
Richard	Connelly	White	Child		1900
Michael	**Connelly**	White	Glass blower	Massachusetts	1900
William	**Connelly**	White	Glass cutter	Massachusetts	1900
Eliazer	**Coon**	White	Glass cutter	New York	1880
Hattie	Coon	White	Wife		1880
Saven	**Coon**	White	Glass cutter	New York	1880
Ellie	Coon	White	Child		1880
Talcort	**Coon**	White	Glass cutter	Massachusetts	1880
Mary	Coon	White	Wife		1880
Willis	**Coon**	White	Glass cutter	New York	1880
Lily	Coon	White	Wife		1880
Effie	Coon	White	Child		1880
Mable	Coon	White	Child		1880
Theodore	**Coones**	White	Day laborer	New York	1900
Mary	Coones	White	Wife		1900
Robert	Coones	White	Child		1900
[daughter]	Coones	White	Child		1900
Mary	**Cornel**	White	Housekeeper		1880
Rosy	**Cornel**	White	Housekeeper		1880
Joseph	**Cousery (?)**	White	Glass melter	Ireland	1880
Josephine	Cousery (?)	White	Wife		1880
Cowen	**Creeder**	White	Glass factory laborer	Germany	1880
George	**Creeder**	White	Glass factory laborer	Massachusetts	1880
John	**Creeder**	White	Glass factory laborer	Germany	1880
Susan	Creeder	White	Wife		1880
Maurice	**Curran**	White	Glass factory laborer	Ireland	1870
Honora	Curran	White	Wife		1870
Thomas	Curran	White	Child		1870
John	Curran	White	Child		1870

Residents of Berkshire Village

First Name	Last Name	Race	Occupation	Origin	Census Year
Mary	Curran	White	Child		1870
Alvan	**Danby**	White	Carpenter	New Hampshire	1850
John	**Daneau**	White	Glass factory laborer	Canada	1870
Harriet	Daneau	White	Wife		1870
Russell	**Darling**	White	Glass factory laborer	New York	1860
Aseneth	David	White	Wife		1850
Calvin	**Davis**	White	Sawyer	New York	1850
Delana	Davis	White	Wife		1850
Wilbur	**Davis**	White	Laborer	Massachusetts	1850
Julia	Davis	White	Wife		1850
Ellen	Davis	White	Child		1850
Laura	Davis	White	Child		1850
Henry	Davis	White	Child		1850
Harriet	Davis	White	Child		1850
John	**Davis**	White	Basket maker; once a slave in NJ, 102 years old	New Jersey	1880
Herbert	**Deaneau**	White	Glass gatherer-shearer	Belgium	1880
Desire	**Dehainant**	White	Glass gatherer	Belgium	1880
Catherine	Dehainant	White	Wife		1880
Desire	Dehainant	White	Child		1880
Martha	Dehainant	White	Child		1880
Humphrey	**Desman**	White	Farm laborer	Ireland	1860
Ellen	Desman	White	Wife	Ireland	1860
John	Desman	White	Child	Ireland	1860
Ellen	Desman	White	Child	Massachusetts	1860
Mary	Desman	White	Child	Massachusetts	1860
James	Desman	White	Child	Massachusetts	1860
Catharine	Desman	White	Child	Massachusetts	1860
Humphrey	**Desmond**	White	Farmer	Ireland	1870
Ellen	Desmond	White	Wife		1870
James	**Desmond**	White	Farm laborer	Massachusetts	1870
Catherine	Desmond	White	Wife		1870
John	**Desmond**	White	Farmer	Ireland	1880
Sarah	Desmond	White	Wife		1880

First Name	Last Name	Race	Occupation	Origin	Census Year
Joseph	Desmond	White	Child		1880
John	Desmond	White	Child		1880
Mary	Desmond	White	Child		1880
James	Desmond	White	Child		1880
Mary	**Desmond**	White	Farmer	Massachusetts	1900
Ellen	Desmond	White	Wife		1900
Humphrey	**Desonel**	White	[unemployed?]	Ireland	1880
Ellen	Desonel	White	Wife		1880
August	**Deulan**	White	[unemployed?]	Belgium	1880
Adelia	Deulan	White	Wife		1880
Tormont	Deulan	White	Child		1880
Octave	Deulan	White	Child		1880
Patrick	**Devlin**	White	Farmer	Ireland	1860
Ann	Devlin	White	Wife	Ireland	1860
Mary J.	Devlin	White	Child	Massachusetts	1860
James J.	Devlin	White	Child	Massachusetts	1860
Eliza	Devlin	White	Child	Massachusetts	1860
John	Devlin	White	Child	Massachusetts	1860
Lucy	Devlin	White	Child	Massachusetts	1860
Thomas	**Dew**	White	Railroad laborer	Ireland	1870
Joseph	**Dolber**	White	Glass factory laborer	Massachusetts	1860
Maria A.	Dolber	White	Wife	Connecticut	1860
Mary J.	Dolber	White	Child	Massachusetts	1860
Sarah M.	Dolber	White	Child	Massachusetts	1860
William	**Donelson**	White	Laborer	New York	1850
C.	**Donovan**	White	Railroad laborer	Ireland	1870
John	**Donovan**	White	Glass factory laborer	Ireland	1870
Thomas	**Donovan**	White	Farmer	Ireland	1870
Mary	Donovan	White	Wife		1870
John	Donovan	White	Child		1870
William	**Donovan**	White	Glass factory laborer	Ireland	1870
Bridget	Donovan	White	Wife		1870
Thomas	**Donovan**	White	Farmer	Ireland	1900

Residents of Berkshire Village

First Name	Last Name	Race	Occupation	Origin	Census Year
William	Donovan	White	Child		1900
Francis	**Donsy**	White	Glass shearer	France	1880
Mary	Donsy	White	Wife		1880
Ameal	Donsy	White	Child		1880
Louie	Donsy	White	Child		1880
Octavy	Donsy	White	Child		1880
John	**Dooley**	White	Day laborer	Massachusetts	1900
Rose	Dooley	White	Wife		1900
Helen	Dooley	White	Child		1900
Grace	Dooley	White	Child		1900
Louise	Dooley	White	Child		1900
John	Dooley	White	Child		1900
Clarence	Dooley	White	Child		1900
Robert	**Dority**	White	Glass factory laborer	Ireland	1880
Darius	**Dorman**	White	Wood chopper	Massachusetts	1860
Mary	Dorman	White	Wife	Massachusetts	1860
Lucinda	Dorman	White	Child	Massachusetts	1860
Lewis	Dorman	White	Child	Massachusetts	1860
Lida	Dorman	White	Child	Massachusetts	1860
Henry	Drum	White	Child		1870
Patrick	**Dulivan**	White	Laborer	Ireland	1850
Ann	Dulivan	White	Wife		1850
Eliza	Dulivan	White	Child		1850
Mary	Dulivan	White	Child		1850
William	**Dulivan**	White	Glass factory laborer	Ireland	1860
Bridget	Dulivan	White	Wife	Ireland	1860
John	Dulivan	White	Glass factory laborer	Ireland	1860
John	**Dulivan**	White	Glass factory laborer	Ireland	1880
Marion	Dulivan	White	Wife		1880
John	**Dulivan**	White	Glass factory laborer	Massachusetts	1880
Andrew	**Dustinhoff**	White	Glass factory laborer	Germany-Possen	1870
Minnie	Dustinhoff	White	Wife		1870
Elizabeth	Dustinhoff	White	Child		1870

First Name	Last Name	Race	Occupation	Origin	Census Year
Charles	**Duval**	White	Painter	New York	1860
Charles	**Egan**	White	Carpenter	Massachusetts	1850
John	**Egan**	White	Railroad laborer	Ireland	1870
Eliza	Egan	White	Wife		1870
John	**Egan**	White	Glass factory laborer	Ireland	1880
Eliza	Egan	White	Wife		1880
John	Egan	White	Child		1880
Donice	Egan	White	Child		1880
Annie	Egan	White	Child		1880
Josey	Egan	White	Child		1880
Daniel	**Egan**	White	Day laborer	Massachusetts	1900
Anna	Egan	White	Wife		1900
John	**Egan**	White	Day laborer	Ireland	1900
Eliza	Egan	White	Wife		1900
William	**Egan**	White	Salesman, groceries	Massachusetts	1900
Marietta	Egan	White	Wife		1900
Emeline	Egan	White	Child		1900
Henry (?)	**Emmalorel**	White	Shoemaker	Germany-Bohemia	1880
Louise	Emmalorel	White	Wife		1880
Hugh	**Fahey**	White	Laborer	Ireland	1850
Nappy	Fahey	White	Child		1850
John	Fahey	White	Child		1850
Patrick	Fahey	White	Child		1850
Bridget	Fahey	White	Child		1850
James	Fahey	White	Child		1850
Ellen	**Fitzgerald**	White	Domestic	Ireland	1860
Michael	**Ford**	White	Glass factory laborer	Ireland	1870
Bridget	Ford	White	Wife		1870
Mary Ann	Ford	White	Child		1870
Bridget	Ford	White	Child		1870
Catherine	Ford	White	Child		1870
Martin	Ford	White	Child		1870
Joseph	**Fournier**	White	Wood chopper	France	1860

Residents of Berkshire Village

First Name	Last Name	Race	Occupation	Origin	Census Year
Mary	Fournier	White	Wife	France	1860
Benjamin	**Franklin**	White	Farm laborer	Massachusetts	1860
John	**Frill**	White	Glass factory laborer	Switzerland	1860
Anna	Frill	White	Wife	Switzerland	1860
John J.	Frill	White	Child	Switzerland	1860
John	**Frilley**	White	Glass factory laborer	Switzerland	1870
Ann	Frilley	White	Wife		1870
John	**Frilley**	White	Glass cutter	Switzerland	1870
Elizabeth	Frilley	White	Child		1870
Arthur	**Frilley**	White	Day laborer	Massachusetts	1900
William	**Fuller**	White	Sand agent	Massachusetts	1850
H (?) [female]	**Fuller**	White	Housekeeper	Massachusetts	1880
Herbert	**Fuller**	White	Bookkeeper	Massachusetts	1880
William	**Fuller**	White	Clerk	Massachusetts	1880
Mary	Fuller	White	Wife		1880
Mary	Fuller	White	Child		1880
Mary	Fuller	White	Wife		1900
Rupell	**Gibbs**	White	Farmer	Massachusetts	1850
Ansel	**Giesnoff**	White	Glass blower	Baden	1860
A.A.	**Gilbert**	White	Teacher at select school	Massachusetts	1860
Mary A.	Gilbert	White	Wife	Massachusetts	1860
A.B.	Gilbert	White	Child	Massachusetts	1860
Julius J.	Gilbert	White	Child	Massachusetts	1860
David	**Gocher**	White	Glass factory laborer	Canada	1860
Charles	**Goshler**	White	Glass factory laborer	Vermont	1880
David	**Gotier**	White	Glass blower	Canada	1870
Margaret	Gotier	White	Wife		1870
Francis	Gotier	White	Child		1870
Mary	Gotier	White	Child		1870
Armenia	Gotier	White	Child		1870
?	**Grady**	White	Wood chopper	Ireland	1860
Mary	Grady	White	Wife	Ireland	1860
Andrew	**Greiner**	White	Glass blower	New York	1870

First Name	Last Name	Race	Occupation	Origin	Census Year
Sarah	Greiner	White	Wife		1870
Rettre	Greiner	White	Child		1870
Jahliel	Greiner	White	Child		1870
Augustus	Greiner	White	Child		1870
Mina	Greiner	White	Child		1870
Carrie	Greiner	White	Child		1870
Bertha	Greiner	White	Child		1870
Sarah	Griswold	White	Child		1870
William	**Gromer**	Black	Laborer	Massachusetts	1850
Maria	Groomer	Mulatto	Child		1870
John	**Grunir**	White	Glass blower	New York	1870
Elizabeth	Grunir	White	Wife		1870
Adam	Grunir	White	Child		1870
Paul	Grunir	White	Child		1870
Cedric	Grunir	White	Child		1870
Luke	Grunir	White	Child		1870
Reuben	Grunir	White	Child		1870
Benjamin	**Gunn**	Black	Laborer	New York	1850
Lucretia	Gunn	Black	Wife		1850
Harriet	Gunn	Black	Child		1850
Selem	Gunn	Black	Child		1850
Eliza	Gunn	Black	Child		1850
John	Gunn	Black	Child		1850
James Jr.	Gunn	White	Child		1850
Thomas	**Hadley**	White	Glass blower	England	1860
Elizabeth	Hadley	White	Wife	England	1860
Elizabeth	Hadley	White	Child	Canada	1860
Sarah	Hadley	White	Child	Canada	1860
Louisa	Hadley	White	Child	New York	1860
William T.	Hadley	White	Child	Massachusetts	1860
Thomas	**Hadley**	White	Glass blower	England	1870
Elizabeth	Hadley	White	Wife		1870
Sarah	Hadley	White	Child		1870
Louisa	Hadley	White	Child		1870
William	Hadley	White	Child		1870
Mary	Hadley	White	Child		1870
Joseph	Hadley	White	Child		1870

Residents of Berkshire Village

First Name	Last Name	Race	Occupation	Origin	Census Year
E (?)	**Hadley**	White	Glass factory laborer	England	1880
Mary	Hadley	White	Wife		1880
Robert	Hadley	White	Glass factory laborer	England	1880
Lewis	Hadley	White	Child		1880
Dennis	**Haley**	White	Farmer	Ireland	1860
John	Haley	White	Farm laborer	Ireland	1860
Ann	Haley	White	Wife	Ireland	1860
Joanna	Haley	White	Child	Ireland	1860
Andrew	Haley	White	Child	Ireland	1860
Betty	**Hall**	White	Child	Massachusetts	1860
Albert	**Hall**	White	Farmer	Rhode Island	1900
Mary	Hall	White	Wife		1900
(?) [female]	**Hambleton**	White	Housekeeper	Massachusetts	1880
Isaac	Hambleton	White	Glass factory laborer	Massachusetts	1880
Ellen	Hambleton	White	Wife		1880
Dulas	**Hambleton**	White	Glass factory laborer	Massachusetts	1880
Melia	Hambleton	White	Wife		1880
Bryant	**Hamilton**	Mulatto	Farm laborer	New Jersey	1860
Hannah	Hamilton	Black	Wife	Connecticut	1860
Emma T.	Hamilton	Mulatto	Child	Massachusetts	1860
Issac D.	Hamilton	Mulatto	Child	Massachusetts	1860
Hannah	**Hamilton**	Black	Wife		1870
Issac	Hamilton	Black	Child		1870
Maria	Hamilton	Black	Child		1870
Joseph	Hamilton	Black	Child		1870
Mary	Hamilton	White	Wife		1880
Aaron	**Hammond**	White	Farmer	Massachusetts	1870
Elizabeth	Hammond	White	Wife		1870
Mary	Hammond	White	Child		1870
Dan	Hammond	White	Child		1870
William	Hammond	White	Child		1870
Frances	Hammond	White	Child		1870
Louisa	Hammond	White	Child		1870
Effie	Hammond	White	Child		1870

First Name	Last Name	Race	Occupation	Origin	Census Year
Lydia	Hammond	White	Child		1870
Nancy	Hammond	White	Child		1870
Thomas	**Hanlin**	White	Teamster	Ireland	1860
Thomas	**Hanlin**	White	Farmer	Ireland	1870
Bridget	Hanlin	White	Wife		1870
William	Hanlin	White	Child		1870
John	Hanlin	White	Child		1870
Julia	Hanlin	White	Child		1870
Ellen	Hanlin	White	Child		1870
Thomas	**Hanlin**	White	Glass factory laborer	Ireland	1880
Bridget	Hanlin	White	Wife		1880
William	**Hanlin**	White	Glass factory laborer	Massachusetts	1880
John	Hanlin	White	Child		1880
Julia	Hanlin	White	Child		1880
Ellen	Hanlin	White	Child		1880
Edward	Hanlin	White	Child		1880
Margaret	Hanlin	White	Child		1880
Patrick	**Hardman**	White	Glass factory laborer	Ireland	1860
Bridget	Hardman	White	Wife	Ireland	1860
Michael	Hardman	White	Child	Massachusetts	1860
Ellen	Hardman	White	Child	Massachusetts	1860
Thomas	Hardman	White	Child	Massachusetts	1860
Maria	Hardman	White	Child	Massachusetts	1860
James	Hardman	White	Child	Massachusetts	1860
Patrick [Jr.]	Hardman	White	Child	Massachusetts	1860
Falean	**Hark**	White	Glass gatherer	Belgium	1880
Charles	**Harris**	White	Farm laborer	Canada	1860
Amelia	Harris	White	Wife	Canada	1860
Charles	Harris	White	Child	Massachusetts	1860
Lewis	Harris	White	Child	Massachusetts	1860
Joseph	Harris	White	Child	Massachusetts	1860
Amelia	Harris	White	Child	Massachusetts	1860
Charles	**Harris**	White	Farmer	Canada	1880
Kate	Harris	White	Wife		1880
John	Harris	White	Child		1880

Residents of Berkshire Village

First Name	Last Name	Race	Occupation	Origin	Census Year
Matilda	Harris	White	Child		1880
Michael	Harris	White	Child		1880
Mary	Harris	White	Child		1880
Sam	**Hazelbaker**	White	Glass cutter	Pennsylvania	1870
Ann	Hazelbaker	White	Wife		1870
Emma	Hazelbaker	White	Child		1870
Elsworth	Hazelbaker	White	Child		1870
Katy	Hazelbaker	White	Child		1870
Lydia	Hazelbaker	White	Child		1870
Catharine	Healy	White	Child	Ireland	1860
Edward	**Hewitt**	White	Glass blower	England	1870
Eliza	Hewitt	White	Wife		1870
Thomas	Hewitt	White	Child		1870
George	Hewitt	White	Child		1870
Sarah	Hewitt	White	Child		1870
William	Hewitt	White	Child		1870
Ann	Hewitt	White	Child		1870
William	**Hewitt**	White	Glass blower	England	1870
Ann	Hewitt	White	Wife		1870
Ann	Hewitt	White	Child		1870
Robert	Hewitt	White	Child		1870
William	Hewitt	White	Child		1870
Edward	Hewitt	White	Child		1870
James	Hewitt	White	Child		1870
Robert	**Hewitt**	White	Glass maker	England	1880
William	Hewitt	White	Child		1880
Edward	Hewitt	White	Child		1880
James	Hewitt	White	Child		1880
Margaret	Hewitt	White	Child		1880
William	**Hewitt**	White	Glass maker	England	1880
Ann	Hewitt	White	Wife		1880
Robert	**Hewitt**	White	Glass gatherer	England	1900
Rosanna	Hewitt	White	Wife		1900
William	**Hewitt**	White	Glass snapper	Massachusetts	1900
George	Hewitt	White	Wife		1900
Mabel	Hewitt	White	Child		1900
Nannie	Hewitt	White	Child		1900

First Name	Last Name	Race	Occupation	Origin	Census Year
Conrad	**Hines**	White	Glass flattener	Germany-Hesse-Darmstadt	1870
Maria	Hines	White	Wife		1870
Mary	Hines	White	Wife		1870
Henry	**Hines**	White	Glass gatherer	Germany-Bremen	1870
Conrad	**Hines**	White	Glass flattener	Germany	1880
Maria	Hines	White	Wife		1880
Agnes	Hines	White	Child		1880
Conrad	Hines	White	Child		1880
Carl	Hines	White	Child		1880
Cheryl	Hines	White	Dressmaker		1880
Conrad	**Hines**	White	Glass flattener	Germany	1900
Mary	Hines	White	Wife		1900
Conrad	**Hines**	White	Glass flattener	Wisconsin	1900
Karl	**Hines**	White	Glass flattener	Massachusetts	1900
Mildred	Hines	White	Wife		1900
Conrad	**Hinse**	White	Glass factory laborer	Switzerland	1860
Elizabeth	Hinse	White	Wife	Switzerland	1860
Conrad J.	Hinse	White	Child	Switzerland	1860
Henry	Hinse	White	Child	Switzerland	1860
Caroline	Hinse	White	Child	Massachusetts	1860
Mary J.	Hinse	White	Child	Massachusetts	1860
Alonzo	**Hoose**	Mulatto	Laborer	Massachusetts	1850
Julia	Hoose	Mulatto	Wife		1850
Marcus	Hoose	Mulatto	Child		1850
Alonzo	Hoose	Mulatto	Child		1850
Mary	Hoose	Mulatto	Child		1850
Algernon	**Hoose**	Black	Farm laborer	Massachusetts	1860
Charles	Hoose	Black	Wife	Massachusetts	1860
Grace	Hoose	Black	Child	Massachusetts	1860
Harriet	Hoose	Black	Child	Massachusetts	1860
Hannah	Hoose	Black	Child	Massachusetts	1860
Alonzo	**Hoose**	Mulatto	Farmer	Massachusetts	1860
Julia	Hoose	Black	Wife	Massachusetts	1860
Alonzo [Jr.]	Hoose	Black	Child	Massachusetts	1860
Frank	Hoose	Black	Child	Massachusetts	1860

Residents of Berkshire Village

First Name	Last Name	Race	Occupation	Origin	Census Year
Alonzo	**Hoose**	Black	Farmer	Massachusetts	1870
Julia	Hoose	Black	Wife		1870
Alonzo	Hoose	Black	Farmer	Massachusetts	1870
Franklin	Hoose	Black	Farmer	Massachusetts	1870
Alonso	**Hoose**	Black	Farmer	Massachusetts	1880
Caroline	Hoose	Black	Wife		1880
Alonso	**Hoose**	Black	Glass factory laborer	Massachusetts	1880
Frank	**Hoose**	Black	Glass factory laborer	Massachusetts	1880
Frank	**Hoose**	Black	Glass factory laborer	Massachusetts	1880
Mary	Hoose	Black	Wife		1880
Frank	Hoose	Black	Child		1880
Charles	Hoose	White	Child		1900
George	**Howard**	White	Butcher	Massachusetts	1860
Emma	Howard	White	Wife	Massachusetts	1860
Cora	Howard	White	Child	Massachusetts	1860
Catherine	**Howe**	White	Teacher at school	Massachusetts	1860
Catharine S.	**Howe**	White	Teacher Select School	Massachusetts	1860
Th.	**Howell**	White	Brick mason	England	1880
Anna	Howell	White	Wife		1880
Ann	**Howell**	White	Wife		1900
Mrs.	**Hungerford**	White	Child		1880
Jacob	**Jackley**	White	Glass factory laborer	Wirtemberg	1860
Mary	Jackley	White	Wife	Wirtemberg	1860
John	Jackley	White	Glass factory laborer	Wirtemberg	1860
Jacob [Jr.]	Jackley	White	Glass factory laborer	Wirtemberg	1860
Agnes	Jackley	White	Child	Wirtemberg	1860
Christian	Jackley	White	Child	Wirtemberg	1860
Thomas	Jackley	White	Child	New York	1860
Mary M.	Jackley	White	Child	New York	1860
Lilly	Jackley	White	Child	Massachusetts	1860

First Name	Last Name	Race	Occupation	Origin	Census Year
Christian	**Jackly**	White	Glass gatherer	Germany-Wirtemberg	1870
Jacob	**Jackly**	White	Glass factory laborer	Germany-Wirtemberg	1870
Maria	Jackly	White	Wife		1870
Lillian	Jackly	White	Child		1870
Thomas	**Jackly**	White	Glass gatherer	New York	1870
Christian	**Jackly**	White	Glass blower	Germany	1880
Lotty	Jackly	White	Wife		1880
Julian	Jackly	White	Child		1880
Thomas	**Jackly**	White	Glass blower	New York	1880
Flora	Jackly	White	Wife		1880
Thomas	Jackly	White	Child		1880
John	Jackly	White	Child		1880
Lillie Belle	Jackly	White	Child		1880
Lillian	Jackman	White	Child		1900
Loretta	Jackman	White	Child		1900
William	Jackman	White	Child		1900
Henry	**Jackson**	Black	Laborer	New York	1850
Lucy	Jackson	Mulatto	Wife		1850
George	Jackson	Mulatto	Child		1850
John	Jackson	Mulatto	Child		1850
Henry	**Jackson**	Black	Farmer	New York	1870
Lucy	Jackson	Mulatto	Wife		1870
Michael	**Jacobs**	White	Glass factory laborer	Ireland	1870
Bridget	Jacobs	White	Wife		1870
Michael	Jacobs	White	Child		1870
Michael	**Jacobs**	White	Glass factory laborer	Ireland	1880
Bridget	Jacobs	White	Wife		1880
Michael	Jacobs	White	Child		1880
Henry	**James**	White	Glass factory laborer	Massachusetts	1880
Lena (?)	James	White	Wife		1880
Henry	James	White	Child		1880
Ira	**Jenks**	White	Farmer	Massachusetts	1850
Harriet	Jenks	White	Wife		1850

Residents of Berkshire Village

First Name	Last Name	Race	Occupation	Origin	Census Year
Albert	Jenks	White	Child		1850
Hellen	Jenks	White	Child		1850
Ransom	Jenks	White	Child		1850
Celestia	Jenks	White	Child		1850
Lydia	Johnson	White	Child		1850
Frank	**Johnson**	White	Glass factory laborer	Massachusetts	1880
Alois	Johnson	White	Wife		1880
Bertha	Johnson	White	Child		1880
Frank	**Johnson**	Black	Mason	Massachusetts	1900
Alice	Johnson	Black	Wife		1900
Abbie	Johnson	Black	Child		1900
Willis	Johnson	Black	Child		1900
Louise	Johnson	Black	Child		1900
Abbie	Johnson	Black	Child		1900
Sylvia	Johnson	Black	Child		1900
Calvin	**Jones**	White	Laborer	Massachusetts	1850
Frances	Jones	White	Wife		1850
Delana	Jones	White	Child		1850
James	Jones	White	Child		1850
William	**Jones**	White	Laborer	Massachusetts	1850
Sasipta	Jones	White	Wife		1850
Henry	**Jones**	Mulatto	Farmer	Massachusetts	1870
Jena	Jones	Mulatto	Wife		1870
Henry	Jones	Mulatto	Child		1870
Lucian	**Jones**	Black	Brickmaker	Massachusetts	1870
Mary	Jones	Black	Wife		1870
Lucian	Jones	Black	Child		1870
Lucian	**Jones**	Black	Glass factory laborer	Massachusetts	1880
Mary	Jones	Black	Wife		1880
Lucius	Jones	Black	Child		1880
George	**Jones**	Black	Day laborer	Massachusetts	1900
Clarence	Jones	Black	Wife		1900
Henry	**Jones**	Black	Day laborer	Massachusetts	1900
Lerna	Jones	Black	Wife		1900
Lucian	**Jones**	Black	Day laborer	Massachusetts	1900
Grace	Jones	Black	Wife		1900

First Name	Last Name	Race	Occupation	Origin	Census Year
Arthur	Jones	Black	Child		1900
Angeline	Jones	Black	Child		1900
Monroe	**Jones**	Black	Farm laborer	Massachusetts	1900
Laurence	**Kearne**	White	Glass packer	Ireland	1870
Catherine	Kearne	White	Wife		1870
Ellen	Kearne	White	Child		1870
Mary	**Keef**	White	Domestic	Ireland	1860
James	**Kelly**	White	Teamster	Ireland	1860
Ann	Kelly	White	Wife	Ireland	1860
Catharine	Kelly	White	Child	Massachusetts	1860
Ann	Kelly	White	Child	Massachusetts	1860
Peter	**Kelly**	White	Blacksmith	Ireland	1860
Mary	Kelly	White	Wife	Ireland	1860
Christopher	Kelly	White	Child	Vermont	1860
Joseph	Kelly	White	Child	Vermont	1860
William	**Kelly**	White	Glass blower	Scotland	1880
Mary	Kelly	White	Wife		1880
Aimey	Kelly	White	Child		1880
Mary	Kelly	White	Child		1880
William	Kelly	White	Child		1880
Annie	**Kelly**	White	Dressmaker		1900
Margaret	Kelly	White	Child		1900
Elizabeth	**Kelly**	White	Milliner		1900
Cecelia	Kelly	White	Child		1900
William	**Kelly**	White	Glass blower	Massachusetts	1900
Mary	Kelly	White	Child		1900
Sidney	**Kelson**	Mulatto	Farm laborer	Massachusetts	1860
Louisa	Kelson	White	Wife	Massachusetts	1860
Nancy M.	Kelson	White	Child	New York	1860
John N.	Kelson	White	Child	New York	1860
Elizabeth	**Kirkenbower**	White	Child	New Jersey	1860
Maria	**Kirkenbower**	White	Child	New Jersey	1860
Henry	**Knight**	White	Lumberman	New York	1850
Catherine	Knight	White	Wife		1850
Frederick	Knight	White	Child		1850
Edgar	Knight	White	Child		1850
George	**Knight**	White	Glass blower	England	1880

Residents of Berkshire Village

First Name	Last Name	Race	Occupation	Origin	Census Year
Hannah	Knight	White	Wife		1880
Ruth	Knight	White	Child		1880
Marian	Knight	White	Child		1880
Samuel	Knight	White	Child		1880
George	Knight	White	Child		1880
George	**Krieger**	White	Farm laborer	Massachusetts	1900
Sussman	**Krieger**	White	Farmer	Germany	1900
George	**Kriger**	White	Glass factory laborer	Germany-Wirtemberg	1870
Susan	Kriger	White	Wife		1870
John	Kriger	White	Child		1870
Emma	Kriger	White	Child		1870
Lena	Kriger	White	Child		1870
George	Kriger	White	Child		1870
William	**Lacart**	White	Glass factory laborer	Canada	1860
Phobe	Lacart	White	Wife	Canada	1860
Phebe A.	Lacart	White	Child	Massachusetts	1860
Almeda	Lacart	White	Child	Massachusetts	1860
Delphine	Lacart	White	Child	Massachusetts	1860
Francis	**Lafever**	White	Glass blower	Belgium	1880
Fellman?	Lafever	White	Wife		1880
Gustav	**Landgraf**	White	Glass blower	New Jersey	1870
Amelia	Landgraf	White	Wife		1870
Amelia	Landgraf	White	Child		1870
James	Landgraf	White	Child		1870
Jena	Landgraf	White	Child		1870
Charles	Landgraf	White	Child		1870
Jennrit	Landgraf	White	Child		1870
Amos	**Lanphier**	White	Farmer	Massachusetts	1850
Naomi	Lanphier	White	Wife		1850
Charles	Lanphier	White	Farmer	Massachusetts	1850
Edward	**Larreau**	White	Glass master shearer	Canada	1870
Lucy	Larreau	White	Wife		1870
Edward	Larreau	White	Child		1870
Lucy	Larreau	White	Child		1870
Charles	Larreau	White	Child		1870

First Name	Last Name	Race	Occupation	Origin	Census Year
Joseph	Larreau	White	Child		1870
Josephine	Larreau	White	Child		1870
Nelly	Larreau	White	Child		1870
Edward	**Larrow**	White	Glass factory laborer	Canada	1860
Lucy	Larrow	White	Wife	Canada	1860
Louisa	Larrow	White	Child	Massachusetts	1860
Adaline	Larrow	White	Child	New York	1860
Edward C.	Larrow	White	Child	Massachusetts	1860
Lucy A.	Larrow	White	Child	Massachusetts	1860
Charles H.	Larrow	White	Child	Massachusetts	1860
Daniel	**Leanson**	White	Stone cutter	Switzerland	1870
Christine	Leanson	White	Wife		1870
George	Leanson	White	Child		1870
John	Leanson	White	Child		1870
Ann	**Leary**	White	Domestic	New York	1860
William	**Lecard**	White	Glass blower	Canada	1870
Phebe	Lecard	White	Wife		1870
Phebe	Lecard	White	Child		1870
Almirda	Lecard	White	Child		1870
Sophie Delphine	Lecard	White	Child		1870
Louisa	Lecard	White	Child		1870
Harriet	Lecard	White	Child		1870
Jessie	Lecard	White	Child		1870
William	Lecard	White	Child		1870
Philomen	**Lechien**	White	Glass blower	Belgium	1880
Francois	Lechien	White	Wife		1880
Selena	Lechien	White	Child		1880
Hosilier (?)	Lechien	White	Child		1880
Louisa	Lechien	White	Child		1880
Francois	Lechien	White	Child		1880
Hauras	Lechien	White	Child		1880
Marie	Lechien	White	Child		1880
Emile	Lechien	White	Child		1880
Daniel	**Lemon**	White	[unemployed?]		1880
Christine	Lemon	White	Wife		1880
Johney	Lemon	White	Child		1880

Residents of Berkshire Village

First Name	Last Name	Race	Occupation	Origin	Census Year
Daniel	**Lennon**	White	Glass factory laborer	Switzerland	1860
Christianna	Lennon	White	Wife	Switzerland	1860
Daniel	Lennon	White	Child	Switzerland	1860
Elizabeth	Lennon	White	Child	Switzerland	1860
John	Lennon	White	Child	Switzerland	1860
Anna	Lennon	White	Child	Switzerland	1860
Rhoda	Lennon	White	Child	New York	1860
Mary	Lennon	White	Child	New York	1860
George	Lennon	White	Child	Massachusetts	1860
Daniel [Jr.]	**Lennon**	White	Glass factory laborer	Switzerland	1860
Hannora	**Lery**	White	Domestic	Ireland	1860
James B.	**Linn**	White	Clerk in store	Massachusetts	1860
John	**Lisett**	White	Glass factory laborer	Ireland	1870
Mary	Lisett	White	Wife		1870
Alice	Lloyd	Mulatto	Child		1850
Amasella	Lloyd	Mulatto	Child		1850
Samuel	**Lloyd**	Mulatto	Laborer	Massachusetts	1850
John	**Loudan**	White	Glass factory laborer	Massachusetts	1860
Mary	**Loudan**	White	Domestic	Massachusetts	1860
Charles	**Lowell**	White	Painter	Massachusetts	1860
Emily	Lowell	White	Wife	Massachusetts	1860
Charles	Lowell	White	Child	Massachusetts	1860
John	**Madden**	White	Laborer	Ireland	1850
Michael	**Madden**	White	Laborer	Ireland	1850
Thomas	**Madden**	White	Laborer	Ireland	1850
Catherine	Madden	White	Wife		1850
Patrick	**Madden**	White	Farm laborer	Ireland	1860
John	**Mageehan**	White	Farm laborer	Ireland	1860
Bridget	Mageehan	White	Wife	Ireland	1860
Mary	Mageehan	White	Child	Massachusetts	1860
Frank	**Mahan**	White	Glass factory laborer	Massachusetts	1880
Katie	Mahan	White	Wife		1880
Jerey	**Mahan**	White	Blacksmith	Canada	1880

First Name	Last Name	Race	Occupation	Origin	Census Year
John	Mahan	White	Glass factory laborer	Massachusetts	1880
Leaser	Mahan	White	Wife		1880
John	Mahan	White	Glass factory laborer	Ireland	1880
Mary	Mahan	White	Wife		1880
William	Mahan	White	Glass factory laborer	Ireland	1880
Katherine	Maharnia	White	Wife		1900
Augustus	Maisa	White	Glass blower	France	1860
Mary	Maisa	White	Wife	France	1860
William	Maisa	White	Child	New York	1860
Louise	Maisa	White	Child	New York	1860
Augustine	Maisa	White	Child	New York	1860
Victor P.	Maisa	White	Child	New York	1860
David	Marks	White	Glass factory laborer	New York	1860
Jerry	Martin	White	Blacksmith	Canada	1870
Louisa	Martin	White	Wife		1870
Jerry Jr.	Martin	White	Child		1870
Herbert	Martin	White	Child		1870
Mary	Martin	White	Wife		1880
Lucy	Martin	White	Child		1880
Herbert	Martin	White	Child		1880
Jessie	Martin	White	Child		1880
Jane	Martin	White	Child		1880
Arthur	Martin	White	Child		1880
Mary	Martin	White	Child		1880
Elisa	Martin	White	Child		1880
John	Martin	White	Bookkeeper	Massachusetts	1880
Eliza	Martin	White	Wife		1880
Charles	Martin	White	Child		1880
William	Martin	White	Glass factory laborer	Ireland	1880
Bridget	Martin	White	Wife		1880
Thomas	Martin	White	Child		1880
John James	Martin	White	Child		1880
Mary	Martin	White	Child		1880

Residents of Berkshire Village

First Name	Last Name	Race	Occupation	Origin	Census Year
Willie	Martin	White	Child		1880
George	Martin	White	Child		1880
Arthur	**Martin**	White	Glass gatherer	Massachusetts	1900
Mary	Martin	White	Wife		1900
Lucy	Martin	White	Child		1900
George	Martin	White	Child		1900
Jerry	**Martin**	White	Blacksmith	Canada	1900
Louise	Martin	White	Wife		1900
Mary	**Mason**	White	Housekeeper	Massachusetts	1880
Ernest	Mason	White	Child		1880
Martin	**McCarty**	White	Glass factory laborer	Ireland	1870
Maria	McCarty	White	Wife		1870
Michael	McCarty	White	Child		1870
Martin	McCarty	White	Child		1870
Maria	McCarty	White	Child		1870
Patrick	McCarty	White	Child		1870
Mary Ann	McCarty	White	Child		1870
Margaret	McCarty	White	Child		1870
Bridget	McCarty	White	Child		1870
Maria	McCarty	White	Child		1900
Martin	McCarty	White	Child		1900
Patrick	**McCarty**	White	Day laborer	Massachusetts	1900
Margaret	McCarty	White	Wife		1900
Henry	McDonald	White	Child		1870
James	**McGowan**	White	Day laborer		1900
John	**McGowan**	White	[unemployed?]	Ireland	1900
Ann	McGowan	White	Wife		1900
John	**McGowan**	White	Day laborer	Massachusetts	1900
Ann	McGowan	White	Wife		1900
Joseph	**McGowan**	White	Day laborer		1900
William	**McGowan**	White	Day laborer		1900
Lawrence	McGowan	White	Child		1900
W.B.	**McLaughlin**	White	Farmer	Maryland	1860
Harriet C.	McLaughlin	White	Wife	Massachusetts	1860
Frank K.	McLaughlin	White	Child	Massachusetts	1860
Frances M.	McLaughlin	White	Child	Massachusetts	1860

First Name	Last Name	Race	Occupation	Origin	Census Year
William	**McLaughlin**	White	Farmer	Merilond [Maryland?]	1880
Harriet	McLaughlin	White	Wife		1880
Frances	McLaughlin	White	Child		1880
William	McLaughlin	White	Child		1880
Susan	McLaughlin	White	Child		1880
Joseph	**Mead**	White	Glass factory laborer	France	1880
Victory	Mead	White	Wife		1880
Joseph	**Mead**	White	Glass factory laborer	New Jersey	1880
Victor	**Mead**	White	Glass factory laborer	New Jersey	1880
Josephine	Mead	White	Wife		1880
Henry	Mead	White	Child		1880
Victoria	Mead	White	Child		1880
Seine (?)	Mead	White	Child		1880
Julius	Mead	White	Child		1880
Flora	Mead	White	Child		1880
Teresa (?)	Mead	White	Child		1880
Bridget	**Meagher**	White	Farmer	Ireland	1900
Katherine	Meagher	White	Wife		1900
George	**Meagher**	White	Day laborer	Massachusetts	1900
John	**Meehan**	White	Glass factory laborer	Ireland	1870
Mary	Meehan	White	Wife		1870
William	**Meehan**	White	Sawmill laborer	Massachusetts	1870
Frank	Meehan	White	Farm laborer	Massachusetts	1870
Mary Anne	Meehan	White	Wife		1870
Catherine	Meehan	White	Child		1870
James	Meehan	White	Child		1870
John	Meehan	White	Child		1870
Elizabeth	Meehan	White	Child		1870
William	**Meehan**	White	Carpenter	Massachusetts	1870
James	**Meehan**	White	Day laborer	Massachusetts	1900
Mary	**Meehan**	White	Wife		1900
Kate	Meehan	White	Child		1900
Stephen	**Minor**	White	Farmer	Massachusetts	1880

Residents of Berkshire Village

First Name	Last Name	Race	Occupation	Origin	Census Year
Ellen	Minor	White	Wife		1880
Hannah	Minor	White	Child		1880
James	**Monyihan**	White	Day laborer	Massachusetts	1900
John	**Monyihan**	White	Day laborer	Ireland	1900
Maggie	Monyihan	White	Wife		1900
Johanna	Monyihan	White	Child		1900
Cornelius	Monyihan	White	Child		1900
Daniel	Monyihan	White	Child		1900
Katherine	Monyihan	White	Child		1900
Patrick	Monyihan	White	Child		1900
Mary	Monyihan	White	Child		1900
John	Monyihan	White	Child		1900
John	**Monyihan**	White	Day laborer	Massachusetts	1900
Joseph	**Monyihan**	White	Day laborer	Massachusetts	1900
William	**Monyihan**	White	Day laborer	Massachusetts	1900
George	**More**	Mulatto	Laborer	Massachusetts	1850
Oliver	**More**	Black	Farmer	Massachusetts	1850
Pamelia	More	Black	Wife		1850
Celia	More	Black	Child		1850
Sylvester	More	Black	Child		1850
Harry	More	Black	Child		1850
Peney	More	Black	Child		1850
Emily	More	Black	Child		1850
Andrew	More	Black	Child		1850
Ishmael	More	Black	Farmer	Massachusetts	1850
Oliver	**Mores**	Black	Farmer	Massachusetts	1870
Mary	Mores	Black	Wife		1870
James	**Morgan**	White	Farmer	Vermont	1850
Harriet	Morgan	White	Wife		1850
Frances	Morgan	White	Child		1850
Julia	Morgan	White	Child		1850
Adelaide	Morgan	White	Child		1850
Michael	**Morrisey**	White	Glass factory laborer	Ireland	1870
Bridget	Morrisey	White	Wife		1870
Thomas	Morrisey	White	Child		1870
John	Morrisey	White	Child		1870

First Name	Last Name	Race	Occupation	Origin	Census Year
Mary	Morrisey	White	Child		1870
Ann	Morrisey	White	Child		1870
William	**Morrisey**	White	Day laborer	Massachusetts	1900
Robert	**Morton**	White	Pauper		1880
Bridget	Mullen	White	Child		1900
John	**Mullen**	White	Glass gatherer	Massachusetts	1900
Peter	Mullen	White	Child		1900
Margaret	Mullen	White	Child		1900
William	**Mullen**	White	Glass gatherer	Massachusetts	1900
William	**Mullin**	White	Mason	Ireland	1870
James	Mullin	White	Mason	Ireland	1870
Maria	Mullin	White	Wife		1870
William	Mullin	White	Child		1870
Thomas	Mullin	White	Child		1870
James	Mullin	White	Child		1870
Joseph	Mullin	White	Child		1870
John	**Murphy**	White	Laborer	Ireland	1850
Lucy	**Murphy**	White	Wife		1850
John	**Murphy**	White	Glass factory laborer	Ireland	1860
Mary	Murphy	White	Wife	Ireland	1860
Patrick	Murphy	White	Child	Ireland	1860
John	Murphy	White	Child	Ireland	1860
Daniel	Murphy	White	Child	Ireland	1860
Catharine	Murphy	White	Child	Massachusetts	1860
Patrick	**Murphy**	White	Farm laborer	Ireland	1860
Ann	Murphy	White	Wife	Ireland	1860
Daniel	Murphy	White	Child	Massachusetts	1860
Heather (?)	Murphy	White	Child	Massachusetts	1860
John	**Murphy**	White	Glass factory laborer	Ireland	1870
Mary	Murphy	White	Wife		1870
John	Murphy	White	Child		1870
Daniel	Murphy	White	Child		1870
Catherine	Murphy	White	Child		1870
Mary	Murphy	White	Child		1870
Patrick	**Murphy**	White	Glass factory laborer	Ireland	1870

Residents of Berkshire Village

First Name	Last Name	Race	Occupation	Origin	Census Year
Ann	Murphy	White	Wife		1870
Daniel	Murphy	White	Child		1870
Kate	Murphy	White	Child		1870
John	Murphy	White	Child		1870
Mary	Murphy	White	Child		1870
Eliza	Murphy	White	Child		1870
Patrick	Murphy	White	Child		1870
Thomas	Murphy	White	Child		1870
Patrick	**Murphy**	White	Glass cutter	Ireland	1870
Joanna	Murphy	White	Wife		1870
John	Murphy	White	Child		1870
John	**Murphy**	White	Glass factory laborer	Ireland	1880
Mary	Murphy	White	Wife		1880
Kate	Murphy	White	Child		1880
Mary	Murphy	White	Child		1880
William	**Murren**	White	Glass factory laborer	Ireland	1880
Bridget	Murren	White	Wife		1880
Maggie	Murren	White	Child		1880
William	Murren	White	Child		1880
Mary	Murren	White	Child		1880
William	**Noonan**	White	Laborer	Ireland	1850
Mary	Noonan	White	Wife		1850
Michael	Noonan	White	Child		1850
Mary	Noonan	White	Child		1850
William	**Noonan**	White	Glass furnace laborer	Ireland	1860
Mary	Noonan	White	Wife	Ireland	1860
Michael	Noonan	White	Child	New York	1860
Mary	Noonan	White	Child	Massachusetts	1860
William	Noonan	White	Child	Massachusetts	1860
Christopher	Noonan	White	Child	Massachusetts	1860
Abigail	Noyes	White	Child		1850
Timothy	**O'Connor**	White	Sawmill laborer	Ireland	1870
Honora	O'Connor	White	Wife		1870
Honora	O'Connor	White	Child		1870
Mary	O'Connor	White	Child		1870

FIRST NAME	LAST NAME	RACE	OCCUPATION	ORIGIN	CENSUS YEAR
Katherine	O'Connor	White	Child		1900
Mary	**Owen**	White	Child	Massachusetts	1880
Mary	Owen	White	Child		1880
William	Owen	White	Child		1880
Lissy	Owen	White	Child		1880
Julia	Owen	White	Child		1880
Aguilla	**Perry**	White	Glass blower	England	1870
Albert	**Perry**	White	Glass gatherer	England	1870
Elizabeth	Perry	White	Wife		1870
Morris	**Perry**	White	Glass gatherer	England	1870
Phyllis	Perry	White	Wife		1870
Angemine	Perry	White	Child		1870
Rose	Perry	White	Child		1870
William	**Perry**	White	Glass factory laborer	England	1870
Calep	**Perry**	White	Glass blower	England	1880
Eva	Perry	White	Servant		1880
Morris	**Perry**	White	Glass blower	England	1880
Philis	Perry	White	Wife		1880
Angemina(?)	Perry	White	Child		1880
Jeremiah	Perry	White	Child		1880
Elizabeth	Perry	White	Child		1880
Thomas [?]	Perry	White	Child		1880
Amos	**Pettibone**	White	Farmer	Connecticut	1850
Lydia	Pettibone	White	Wife		1850
Harvey	Pettibone	White	Wagonmaker	Massachusetts	1850
Frances	Pettibone	White	Child		1850
Charles	Pettibone	White	Farmer	Massachusetts	1850
Bishop	Pettibone	White	Child		1850
Sarah	Pettibone	White	Child		1850
Cecil	Pettibone	White	Child		1850
John	**Pettibone**	White	Farmer	Massachusetts	1850
Elvira	Pettibone	White	Wife		1850
Sarah	Pettibone	White	Child		1850
Elizabeth	Petty	White	School teacher	Massachusetts	1870
Henry	**Pines**	White	Glass gatherer	Germany-Bohemia	1870
John	**Pinketh**	White	Glass gatherer	England	1870

Residents of Berkshire Village

First Name	Last Name	Race	Occupation	Origin	Census Year
Emma	Pinketh	White	Wife		1870
William	Pinketh	White	Child		1870
Philip	**Porter**	White	Carpenter	Massachusetts	1850
Martha	Porter	White	Wife		1850
Martha	Porter	White	Child		1850
Philip	Porter	White	Child		1850
Selden	Porter	White	Child		1850
Sarah	Porter	White	Child		1850
Nelson	Porter	White	Child		1850
Milton	Porter	White	Child		1850
Philip	**Porter**	White	Carpenter	Massachusetts	1860
Martha	Porter	White	Wife	Massachusetts	1860
Martha [Jr.]	Porter	White	Child	Massachusetts	1860
Philip	Porter	White	Farm laborer	Massachusetts	1860
Selden	Porter	White	Child	Massachusetts	1860
Sarah	Porter	White	Child	Massachusetts	1860
Nelson	Porter	White	Child	Massachusetts	1860
William	Porter	White	Child	Massachusetts	1860
William	Porter	White	Child	Massachusetts	1860
Martha	Porter	White	Child		1870
Milton	**Porter**	White	Farmer	Massachusetts	1870
Willie	Porter	White	Wife		1870
Merdson	Porter	White	Child		1870
Philip	**Porter**	White	Farmer	Massachusetts	1870
Martha	Porter	White	Wife		1870
Martha	Porter	White	Child		1870
Seldon	**Porter**	White	Carpenter	Massachusetts	1870
Martha	**Porter**	White	Child		1880
Martha	Porter	White	Child		1880
Milton	**Porter**	White	Glass factory laborer	Massachusetts	1880
Phillip	**Porter**	White	Glass factory laborer	Massachusetts	1880
Jane	Porter	White	Wife		1880
Jennie	Porter	White	Child		1880
William	Porter	White	Child		1880
Philip	Porter	White	Child		1880
George	Porter	White	Child		1880

First Name	Last Name	Race	Occupation	Origin	Census Year
Seldon	**Porter**	White	Glass factory laborer	Massachusetts	1880
William	**Porter**	White	Glass factory laborer	Massachusetts	1880
George	**Porter**	White	Day laborer	Massachusetts	1900
Jennie	**Porter**	White	Rag sorter at paper mill	Massachusetts	1900
Rhoda	Potter	Black	Child		1870
Sloan	**Powell**	White	Farmer	Massachusetts	1870
Ruth	Powell	White	Wife		1870
George	Powell	White	Farmer		1870
Sarah	Powell	White	Child		1870
Flora	Powell	White	Child		1870
Thomas	**Powers**	White	Glass factory laborer	Ireland	1860
Mary	Powers	White	Wife	Ireland	1860
Mary	Powers	White	Child	Massachusetts	1860
Benjamin P.	**Pratt**	White	Farm laborer	Massachusetts	1860
Esther P.	Pratt	White	Wife	Massachusetts	1860
Fanny M.	Pratt	White	Child	Massachusetts	1860
Thomas	**Pye**	White	Glass blower	England	1870
Mary	Pye	White	Wife		1870
Bertram	Pye	White	Child		1870
James	**Raybold**	White	Glass pot maker	Massachusetts	1860
Anphella	Raybold	White	Wife	Massachusetts	1860
Howard E.	Raybold	White	Child	Massachusetts	1860
James	**Raybold**	White	Glass factory overseer	England	1860
Elizabeth	Raybold	White	Wife	Massachusetts	1860
Caroline A.	Raybold	White	Child	Massachusetts	1860
John H.	Raybold	White	Child	New York	1860
Julia E.	Raybold	White	Child	New York	1860
Estella A.	Raybold	White	Child	Massachusetts	1860
Elizabeth	**Raybold**	White	Wife		1870
John	Raybold	White	Child		1870
Estella	Raybold	White	Child		1870
Walter	**Raybold**	White	Glass factory superintendent	Massachusetts	1870

Residents of Berkshire Village

First Name	Last Name	Race	Occupation	Origin	Census Year
Harriet	Raybold	White	Wife		1870
William	Raybold	White	Child		1870
Fanny	Raybold	White	Child		1870
Walter	Raybold	White	Child		1870
John	**Raybold**	White	Glass manufacturer	Massachusetts	1880
Elizabeth	Raybold	White	Wife		1880
Fred	**Reading**	White	Glass cutter	New York	1870
Mary	Reading	White	Wife		1870
John	Reading	White	Child		1870
Jessie	Reading	White	Child		1870
Fred	**Reading**	White	Glass cutter	New York	1880
Mary	Reading	White	Wife		1880
Jason	Reading	White	Child		1880
Louisa	Reading	White	Child		1880
Joseph	**Recor**	White	Glass factory laborer	Canada	1870
Margaret	Recor	White	Wife		1870
John	Recor	White	Child		1870
Julian	Recor	White	Child		1870
James	Recor	White	Child		1870
Elizabeth	Recor	White	Child		1870
Frederick	**Redding**	White	Glass cutter	New York	1900
Mary	Redding	White	Wife		1900
John	Reding	White	Child		1880
Agile	**Richard**	Black	Glass factory laborer	Massachusetts	1880
Fre'd (?)	**Richard**	White	Glass factory laborer	Connecticut	1880
Grace	Richard	White	Wife		1880
Sarah	Richard	White	Child		1880
Arthur	Richard	White	Child		1880
Alexander	**Richards**	White	Glass factory laborer	New York	1860
Saphronia	Richards	White	Wife	Massachusetts	1860
Roxanna	**Richards**	Black	Wife	Massachusetts	1860
Angelina	Richards	Black	Child	Massachusetts	1860
May	Richards	Black	Child	Massachusetts	1860

First Name	Last Name	Race	Occupation	Origin	Census Year
Benjamin Franklin	Richards	Black	Farm laborer	Massachusetts	1860
Frederick	**Richards**	Black	Farmer	Massachusetts	1870
Triphina	Richards	Black	Wife		1870
Hannah	Richards	Black	Child		1870
Ardeline	Richards	Black	Child		1870
Theodore	**Richards**	Black	Glass factory laborer	Massachusetts	1870
Angelina	Richards	Black	Wife		1870
Julia	Richards	Mulatto	Child		1870
Henry	Richards	Mulatto	Child		1870
James	Richards	Mulatto	Child		1870
Thomas	**Richardson**	White	Glass blower	England	1870
Louisa	Richardson	White	Wife		1870
John	**Richardson**	White	Glass blower	England	1900
Venus	Richardson	White	Wife		1900
Thomas	Richardson	White	Child		1900
Charles	**Riding**	White	Glass cutter	New York	1860
Owen	**Riley**	White	Laborer	Ireland	1850
Sarah	Riley	White	Wife		1850
Owen	Riley	White	Child		1850
Edward	**Roberts**	Black	Laborer	Massachusetts	1850
Sarah	Roberts	Black	Wife		1850
Lucy	Roberts	Black	Child		1850
Martha	Roberts	Black	Child		1850
Frederick	**Roberts**	Black	Laborer	Massachusetts	1850
Mary	Roberts	Black	Wife		1850
Genette	Roberts	Black	Child		1850
William	Roberts	Black	Child		1850
Sylva	Roberts	Black	Child		1850
George	Roberts	Black	Child		1850
Theodore	Roberts	Black	Child		1850
Sarah	**Roberts**	Black	Child		1850
Alexander	Roberts	Black	Child		1850
Lucy	Roberts	Black	Child	Massachusetts	1860
Frederick	**Roberts**	White	Glass factory laborer	Massachusetts	1870
Abbie	Roberts	White	Wife		1870

Residents of Berkshire Village

First Name	Last Name	Race	Occupation	Origin	Census Year
Sylvia	Roberts	White	Child		1870
Huldah	Roberts	White	Child		1870
Alice	Roberts	White	Child		1870
Eugene	Roberts	Black	Child		1870
Lewis	**Roberts**	White	Glass Glass factory laborer	Canada	1870
Celeste	Roberts	White	Wife		1870
William	Roberts	White	Child		1870
Celeste	Roberts	White	Child		1870
Abigail	**Roberts**	Black	Housekeeper		1880
Alonso	Roberts	Black	Child		1880
Maria	Robinson	Mulatto	Child		1850
William	**Rogenia**	White	Glass factory laborer	Belgium	1880
Samuel	**Rogerson**	White	Glass blower	England	1870
Albert	**Rogerson**	White	Day laborer	Massachusetts	1900
Thomas	Rogerson	White	Child		1900
Charlotte	Rogerson	White	Child		1900
George	**Rogerson**	White	Glass blower	England	1900
Sophia	Rogerson	White	Wife		1900
Alice	Rogerson	White	Child		1900
Josephine	**Rose**	White	Farmer	New Jersey	1900
Robert	**Russell**	White	Glass factory laborer	Ireland	1870
Ann	Russell	White	Wife		1870
Jane	Russell	White	Child		1870
William	Russell	White	Child		1870
Michael	**Ryan**	White	Laborer	Ireland	1850
Margaret	Ryan	White	Wife		1850
John	**Savey**	White	Laborer	Ireland	1850
Albert	Sears	White	Child		1870
Russ	Semfason	White	Child		1880
Rodney	**Shade**	White	Glass blower	New Jersey	1870
Elizabeth	Shade	White	Wife		1870
Jessie	Shade	White	Child		1870
Samuel B.	**Shaw**	White	Clergy, Episcopal	Rhode Island	1860
Caroline F.	Shaw	White	Wife	South Carolina	1860

First Name	Last Name	Race	Occupation	Origin	Census Year
Anna	Shaw	White	Child	Massachusetts	1860
Abby	Shaw	White	Child	Massachusetts	1860
John	**Shaw**	White	[unemployed?]		1870
Sarah	Shaw	White	Wife		1870
Fanny	**Shepard**	Black	Wife		1850
Zechariah	Shepard	Black	Laborer	Vermont	1850
Abeal	**Shepardson**	White	Farmer	Massachusetts	1850
Maria	Shepardson	White	Wife		1850
Harriet	Shepardson	White	Child		1850
Maria	Shepardson	White	Child		1850
James	**Shepardson**	White	Farmer	Massachusetts	1850
Lydia	Shepardson	White	Wife		1850
Truman	Shepardson	White	Child		1850
Lewis	**Shepardson**	White	Farmer	Massachusetts	1850
Lois	Shepardson	White	Wife		1850
Augusta	Shepardson	White	Child		1870
Jasper	**Shiregt (?)**	White	Glass blower	England	1880
Louisa	Shiregt (?)	White	Wife		1880
Genetta	Shiregt (?)	White	Child		1880
Jasper	Shiregt (?)	White	Child		1880
Lottia	Shiregt (?)	White	Child		1880
August	**Silgious**	White	Glass gatherer	Belgium	1880
John	Silk	White	Laborer	Ireland	1850
Thomas	Silk	White	Laborer	Ireland	1850
John	**Silk**	White	Wood chopper	Ireland	1860
Hannora	Silk	White	Wife	Ireland	1860
Catharine	Silk	White	Child	Ireland	1860
Hannora	Silk	White	Child	Massachusetts	1860
John	Silk	White	Child	Massachusetts	1860
James	Silk	White	Child	Massachusetts	1860
Michael	**Silk**	White	Farm laborer	Ireland	1860
Thomas	**Silk**	White	Farm laborer	Ireland	1860
Bridget	Silk	White	Wife	Ireland	1860
Thomas [Jr.]	Silk	White	Child	Massachusetts	1860
Mary	Silk	White	Child	Massachusetts	1860
Florence	Silk	White	Child		1870
Honora	**Silk**	White	Child		1870

Residents of Berkshire Village

First Name	Last Name	Race	Occupation	Origin	Census Year
Honora	Silk	White	Wife		1870
Ellen	Silk	White	Child		1870
John	Silk	White	Child		1870
James	Silk	White	Child		1870
Mary	Silk	White	Domestic		1870
John	**Silk**	White	Glass factory laborer	Massachusetts	1880
Hannah	Silk	White	Wife		1880
Jonas	**Silk**	White	Glass factory laborer	Massachusetts	1880
Hannah	**Silk**	White	Farmer	Ireland	1900
[son]	**Silk**	White	Farm laborer	Massachusetts	1900
William	**Skull**	White	Glass blower	New Jersey	1870
Olive	Skull	White	Wife		1870
Catherine	Skull	White	Child		1870
Jane	Skull	White	Child		1870
Rebecca	Skull	White	Child		1870
E. Kirby	Skull	White	Child		1870
Anna	Skull	White	Child		1870
Ephraim	**Slade**	White	Laborer	New Hampshire	1850
William	**Smith**	White	Blacksmith	Massachusetts	1860
Susan	Smith	White	Wife	New York	1860
William R.	Smith	White	Child	Massachusetts	1860
William R.	**Smith**	White	Teacher, music	Massachusetts	1860
John	**Smith**	White	Glass factory laborer	Massachusetts	1870
John	**Souden**	White	Glass blower's asst	England	1870
Harriet	Souden	White	Wife		1870
Jennie	Souden	White	Child		1870
Jessie	Souden	White	Child		1870
Fred	Souden	White	Child		1870
Ida	Souden	White	Child		1870
Edwin	Sparhawk	White	Child		1850
George	**Spencer**	White	Glass factory laborer	England	1860
Sophia	Spencer	White	Wife	New York	1860
Henry	**Squire**	White	Farmer	Vermont	1850

First Name	Last Name	Race	Occupation	Origin	Census Year
Joseph	**Stanley**	White	Glass blower	England	1870
Julia	Stanley	White	Wife		1870
May	Stanley	White	Child		1870
John	**Stephens**	White	Carpenter	Massachusetts	1870
Priscilla	Stephens	White	Wife		1870
John	**Stephens**	White	Carpenter	Massachusetts	1900
Priscilla	Stephens	White	Wife		1900
Helen	Stephens	White	Child		1900
John	**Stevens**	White	Carpenter	Massachusetts	1880
Priscilla	Stevens	White	Wife		1880
John	**Stingar**	White	Glass blower	Maryland	1860
Panne	**Stone**	White	Glass flattener	England	1870
Alice	Stone	White	Wife		1870
Lucy	Stone	White	Child		1870
Eliza	Stone	White	Child		1870
Frank	**Stone**	White	Day laborer	Canada	1900
Mary	Stone	White	Wife		1900
George	Stone	White	Child		1900
Edwin	**Sturdevant**	White	Laborer	Massachusetts	1850
Nora	**Sulivan**	White	Servant	Ireland	1880
David	**Sullivan**	White	Farmer	Ireland	1870
Catherine	Sullivan	White	Wife		1870
John	Sullivan	White	Child		1870
Jerry	**Sullivan**	White	Glass blower	Ireland	1870
Mary	Sullivan	White	Wife		1870
John	Sullivan	White	Child		1870
Honora	Sullivan	White	Child		1870
Mary	Sullivan	White	Child		1870
David	**Sullivan**	White	Clerk	Ireland	1880
James	Sullivan	White	Wife		1880
Ellen	Sullivan	White	Child		1880
Mary	Sullivan	White	Child		1880
Bridget	Sullivan	White	Child		1880
Fedely	Sullivan	White	Child		1880
Mike	Sullivan	White	Child		1880
Michael	**Sullivan**	White	Glass factory laborer	Ireland	1880

Residents of Berkshire Village

First Name	Last Name	Race	Occupation	Origin	Census Year
Catherine	Sullivan	White	Wife		1880
John	**Sullivan**	White	Glass gatherer	Pennsylvania	1900
Mary	Sullivan	White	Wife		1900
George	**Swan**	White	Glass gatherer	England	1870
Louisa	Swan	White	Wife		1870
Mary Ann	Swan	White	Child		1870
George	**Swan**	White	Glass factory laborer	Scotland	1880
Mary Ann	Swan	White	Wife		1880
Eliza	Swan	White	Child		1880
Louisa	Swan	White	Child		1880
Lilly	Swan	White	Child		1880
Annie	Swan	White	Child		1880
George	Swan	White	Child		1880
Julia	Sweet	White	Child	Massachusetts	1860
William	**Tabron**	White	Glass factory laborer	England	1870
Jane	Tabron	White	Wife		1870
Thomas	Tabron	White	Child		1870
George	Tabron	White	Child		1870
Elizabeth	Tabron	White	Child		1870
William	Tabron	White	Child		1870
John	Tabron	White	Child		1870
Henry	**Tankard**	Black	Glass factory laborer	New York	1870
Mary	Tankard	Black	Wife		1870
Sophronia	Tankard	Black	Child		1870
Martha	Tankard	Black	Child		1870
Carrie	Tankard	Black	Child		1870
Matthew	**Tankerd**	Mulatto	Glass factory laborer	Vermont	1860
Elizabeth	Tankerd	Mulatto	Wife	New York	1860
William H.	Tankerd	Mulatto	Child	New York	1860
Mary E.	Tankerd	Mulatto	Child	New York	1860
William	**Tankerd**	White	Glass factory laborer	New York	1860
Lan (?)	**Tassece (?)**	White	Glass factory laborer	Belgium	1880
Eliza	Tassece (?)	White	Wife		1880

First Name	Last Name	Race	Occupation	Origin	Census Year
William	Tassece (?)	White	Child		1880
Jessie (?)	Tassece (?)	White	Child		1880
Daniel	**Thomas**	White	Farm laborer	Massachusetts	1860
Daniel	**Thomas**	White	Farm laborer	Massachusetts	1860
Hannah	Thomas	White	Wife	New York	1860
Ann Murgeson	Thornber	White	Child		1850
Albert	**Tolman**	White	Teacher at school	Massachusetts	1860
Jane A.	Tolman	White	Wife	Massachusetts	1860
Carlton T.	Tolman	White	Child	Massachusetts	1860
Albert H.	Tolman	White	Child	Massachusetts	1860
William	Tolman	White	Child	Massachusetts	1860
George	Tolman	White	Child	Massachusetts	1860
S (?)	**Toshie**	White	Glass cutter	Belgium	1880
Eliza	Toshie	White	Wife		1880
Willis	Toshie	White	Child		1880
Elizabeth	Toshie	White	Child		1880
Jarvis	**Town**	White	Carpenter	Massachusetts	1860
John	**Tucker**	Mulatto	Laborer	Massachusetts	1850
Eliza	Tucker	Mulatto	Wife		1850
Charles	Tucker	Mulatto	Laborer	Massachusetts	1850
Nahum	Tucker	Mulatto	Laborer	Massachusetts	1850
Oliver	Tucker	Mulatto	Child		1850
Jane	Tucker	Mulatto	Wife	Massachusetts	1860
William	Tucker	Mulatto	Child	Massachusetts	1860
Adeline	Tucker	Mulatto	Child		1870
Harriet	Tucker	Mulatto	Child		1870
John	**Tucker**	Mulatto	Child		1870
Thomas	**Tulley**	White	Glass factory laborer	Ireland	1880
Mary	Tulley	White	Wife		1880
Pat	Tulley	White	Child		1880
Lucy	Tulley	White	Child		1880
Martin	Tulley	White	Child		1880
William	**Tussier**	White	Glass melter	Belgium	1880
Franklin	**Tyrrel**	White	Farmer	Massachusetts	1850
Anne	Tyrrel	White	Wife		1850
Samuel	Tyrrel	White	Child		1850

Residents of Berkshire Village

First Name	Last Name	Race	Occupation	Origin	Census Year
Alonzo	Tyrrel	White	Child		1850
Levi	Tyrrel	White	Child		1850
Louise	**Valken (?)**	White	Child		1900
Deanthe	VanAllstyne	Black	Child		1850
William	**VanAllstyne**	Black	Laborer	New York	1850
Nancy	VanAllstyne	Black	Wife		1850
Wallace	VanAllstyne	Black	Laborer	Massachusetts	1850
Peter	**Vandema**	White	Glass gatherer	Belgium	1880
Artami	Vandema	White	Wife		1880
Laura	Vandema	White	Child		1880
Carlo	Vandema	White	Child		1880
Iner	Vandema	White	Child		1880
Frances	Vandema	White	Child		1880
Mary	Vandema	White	Child		1880
Victoria	Vandema	White	Child		1880
Simon	**Wager**	White	Day laborer	New York	1900
Elsie	Wager	White	Wife		1900
Minnie	Wager	White	Child		1900
Louis	Wager	White	Child		1900
John	**Waldrin**	White	Clergyman	New York	1900
Bessie	Waldrin	White	Wife		1900
La Pricelle Joan	Waldrin	White	Child		1900
Ina	Waldrin	White	Child		1900
John	**Walker**	White	Teamster	Massachusetts	1860
Harriet C.	Walker	White	Wife	Massachusetts	1860
Mary	Walker	White	Child	Massachusetts	1860
Elizabeth	Walker	White	Child	Massachusetts	1860
Bosalou	**Wallet**	White	Glass gatherer	Belgium	1880
C.R.	**Washburn**	White	Bookkeeper	Massachusetts	1870
Carrie	Washburn	White	Wife		1870
Ella	Washburn	White	Child		1870
Luther	**Washburn**	White	Glass factory overseer	Massachusetts	1870
Laura	Washburn	White	Wife		1870
Peter	**Weard**	White	Glass factory laborer	Ireland	1880
Catherine	Weard	White	Wife		1880

First Name	Last Name	Race	Occupation	Origin	Census Year
Josey	Weard	White	Child		1880
Joseph	Webb	White	Laborer	England	1850
Jacob	**Webb**	White	Glass blower	New Jersey	1860
Cyrus	**Werden**	White	Carpenter	Massachusetts	1850
Juliette	Werden	White	Wife		1850
Francis	Werden	White	Child		1850
William	**Werden**	White	Merchant	Massachusetts	1850
Jacob	**Wheeler**	White	Farmer	New York	1850
Mary	Wheeler	White	Wife		1850
Alisa	Wheeler	White	Child		1850
Lewis	**Wheeler**	White	Clerk in store	Massachusetts	1870
Julia	Wheeler	White	Wife		1870
Arthur	Wheeler	White	Child		1870
Maria	Wheeler	White	Child		1870
Mary	**Wheeler**	White	Wife		1870
Harriet	Wheeler	White	Child		1870
Francis	Wheeler	White	Child		1870
Emily	Wheeler	White	Child		1870
Adelbert	Wheeler	White	Child		1870
Delbert	**Wheeler**	White	Glass factory laborer	Massachusetts	1880
Francis	**Wheeler**	White	Glass factory laborer	Massachusetts	1880
Gerttrine	Wheeler	White	Wife		1880
George	Wheeler	White	Child		1880
Mary	**Wheeler**	White	Housekeeper	Massachusetts	1880
Mary	**Wheeler**	White	Child		1900
Stephen	**Whipple**	White	Farmer	Massachusetts	1850
Lucy	Whipple	White	Wife		1850
Gertrude	Whipple	White	Child		1850
Walter	Whipple	White	Child		1850
Phila	Whipple	White	Child		1850
Benjamin	**Whipple**	White	Glass box maker	Massachusetts	1860
Sally	Whipple	White	Wife	Massachusetts	1860
Stephen T.	**Whipple**	White	Farmer and manager	Massachusetts	1860
Phebe J.	Whipple	White	Wife	Massachusetts	1860

Residents of Berkshire Village

First Name	Last Name	Race	Occupation	Origin	Census Year
Gertrude M.	Whipple	White	Child	Massachusetts	1860
Clara J.	Whipple	White	Child	Massachusetts	1860
Kate U.	Whipple	White	Child	Massachusetts	1860
Truman	**Whipple**	White	Farmer	Massachusetts	1860
Emma J.	Whipple	White	Wife	Vermont	1860
Joseph S.	Whipple	White	Child	Massachusetts	1860
Esther E.	Whipple	White	Child	Massachusetts	1860
Nathan	**Whitcomb**	White	Glass factory laborer	Massachusetts	1880
Sam	**Whitcomb**	White	Glass factory laborer	Massachusetts	1880
Candace	Whitcomb	White	Wife		1880
Ebbin	Whitcomb	White	Child		1880
George	Whitcomb	White	Child		1880
Eliot	Whitcomb	White	Child		1880
John	**Whittrick**	White	Glass blower	New Hampshire	1860
Daniel	Wilkinson	White	Laborer	England	1850
Jonas	**Williams**	White	Depot master	New York	1880
Emeline	Williams	White	Wife		1880
Charles	**Williams**	Black	Farmer	Massachusetts	1900
Angeline	Williams	Black	Wife		1900
Mabel	Williams	Black	Child		1900
Agnes	Williams	Black	Child		1900
James	**Williams**	White	Station agent	New York	1900
Emeline	Williams	White	Wife		1900
Daniel	**Willis**	White	Glass blower	England	1870
Mary	Willis	White	Wife		1870
John	Willis	White	Child		1870
Isabelle	Willis	White	Child		1870
Mary	Willis	White	Child		1870
Margaret	Willis	White	Child		1870
Anne	Willis	White	Child		1870
Augustus	Willis	White	Child		1870
Richard	**Winner**	White	Glass blower	Pennsylvania	1860
Julia A.	Winner	White	Wife	Baden	1860
George W.	Winner	White	Child	New York	1860
Julia	Winner	White	Child	New York	1860
Richard H.	Winner	White	Child	New York	1860

First Name	Last Name	Race	Occupation	Origin	Census Year
Albert F.	Winner	White	Child	Massachusetts	1860
Emma S.	Winner	White	Child	Massachusetts	1860
James	**Winslow**	White	Glass factory laborer	Massachusetts	1870
Cornelius	**Wood**	White	Farmer	Massachusetts	1880
George	**Wood**	White	Farmer	Massachusetts	1880
Harriet	Wood	White	Wife		1880
Oliver	Wood	White	Child		1880
William	**Yeomans**	White	Glass flattener	England	1880
Anna	Yeomans	White	Wife		1880
Sarah	Yeomans	White	Child		1880
Lucy	Yeomans	White	Child		1880
Eliza	Yeomans	White	Child		1880
Annie	Yeomans	White	Child		1880

Appendix III

1886 Fire Insurance Map Descriptions

No. 9000
Berkshire Glass Works,
Berkshire, Lanesboro, Mass.

Owned—By Berkshire Glass Company.
Goods—Plate glass.
Stock—Sand, lime, soda ash, etc.
Capacity—Employ 110 hands.

Power—None.
Exposure—Frame dwellings.
Surveyed—December 9, 1886, by L.M.B.

Description.

No. 1—**Main Building**—Height—One high story. Size———Walls—frame. Roofs—shingle. Cornice—wood. Scuttle—none. Floors—earth. Ceilings—open to roof. Stairs—none. Elevator—none.
Occupation—Gas heated furnace and blowing, batch or mixing room in addition.
No. 2—**Plate Casting Factory**—One high story, frame, shingle and gravel roof, earth floor; furnace, annealing ovens, cutting and packing.
No. 3—**Factory**—One story, frame, shingle roof, earth floor, used only for storage.
No. 4—**Straightening Factory**—One story, frame, shingle and slate roofs, furnace for straightening plates.

No. 5—CUTTING & PACKING HOUSE—One story and attic, frame, box plastered, shingle roof. First story, cutting, packing, and store glass. Attic, store hay for packing.

No. 6—GAS HOUSE—One story and basement, frame, shingle roof, contain retorts and small boiler.

No. 7—BRICK POT HOUSE—Two stories, brick, slate roof, brick boiler house adjoining.

No. 8—FRAME POT HOUSE—Two stories, frame, box plastered, shingle roof.

No. 9—NEW CUTTING SHOP—One story, frame, brick basement, shingle roof, used for storage.

SPECIAL FEATURES.

Heating—Coal stoves set on zinc.

Lighting—Kerosene oil.

Watchmen—None when not running, from 7 to 20 men at work when running.

Furnaces & Ovens—are of brick, in good condition, and clear from wood work.

Drying—of sand is done in **No. 1**, mixing room around a stove enclosed in sheet iron.

Oils—None.

Waste—None.

Hours of work—Day and night.

Boilers—Small upright in gas house. Tubular in addition to **No. 7** not used.

FIRE APPLIANCES.

Fire Pump—None.

Vertical Pipe—None.

Tank—None.

Hydrants—None.

Hose—None.

Sprinklers—None.

Casks and Buckets—None.

Steam Jets—None.

Extinguishers—Two Babcock, and hand grenades.

Lightning Rods—None.

Ladders—Movable.

Auxilary Aid and Signals—None.

CHARACTER.

Buildings of light frame construction, in fair repair, near together and exposing each other. Furnaces and ovens in good condition. No fire appliances. Appear to be doing quite an extensive business which is under good management.

Warshaw Collection, Archives Center, National Museum of American History, Smithsonian Institution.

Notes

Introduction

1. J.E.A. Smith, *History of Pittsfield, (Berkshire County) Massachusetts, From the Year 1800 to the Year 1876* (Springfield, MA: C.W. Bryan & Co., 1876), 545; *History of Berkshire County, Massachusetts with Biographical Sketches of its Prominent Men* (New York: J.B. Beers, 1885), 422.

Sand

2. Ed Kirby, "Preliminary Report on Glass Sand Formation in the Lenox, Massachusetts Region" (unpublished paper, Lenox [Mass.] Historical Society, September 18, 1995), 7.

3. Edward Hitchcock, *Report on the Geology, Mineralogy, Botany, and Zoology of Massachusetts* (Amherst, MA, 1835), 33–34; "Art. IX.—Extracts from the Minutes of the Lyceum of Natural History," *New York Journal of Medicine and Collateral* 10 (May 1848): 340.

4. Joseph D. Weeks, "Glass Materials," in Albert Williams Jr., *Mineral Resources of the United States* (Washington, D.C.: Government Printing Office, 1883–84), 960–61; Charles Fettke, "Special Articles: The Glass Sands of Pennsylvania," *Science* 48 (July 26, 1918): 100. Sand formed by erosion has rounded grains that do not melt evenly.

5. Kirby, "Preliminary Report," 2–7.

6. "Art. IX.—Extracts," 341.

7. William G. Harding, "Glass Manufacture in the Berkshires," *Berkshire Historical and Scientific Society* (Pittsfield, MA: Sun Printing Co., 1894), 39; *History of the County of Berkshire, Massachusetts, in Two Parts* (Pittsfield, MA: Samuel Bush, 1829), 393.

8. Edward Hitchcock, "Report on the Geology of Massachusetts during the years 1830 and 1831," *American Journal of Science* 22 (July 1832): 40.

9. Thomas Gaffield, *Glass Journal* 3, Thomas Gaffield Papers (MC 139), Institute Archives and Special Collections, Massachusetts Institute of Technology (MIT) Libraries, Cambridge, Massachusetts, 115.

10. Ibid., vol. 1, 9–10.

11. "Art. IX.—Extracts," 341.

12. Ibid.; Gaffield, *Glass Journal*, vol. 3, 115.

13. "Knowledge Is Power," *The Friend*, October 25, 1856, 49; "Art. IX.—Extracts," 341.

14. Quoted in Joseph D. Weeks, *Report on the Manufacture of Glass* (Washington, D.C.: United States Census Office, 1884), 26.

15. See William B. Browne, "Over Pathways of the Past," *North Adams Transcript*, June 11, 1938, 10. This article mentions a diary kept by William Fuller, a resident of Berkshire Village who was the agent for the sand mines in Cheshire. We have been unable to locate this diary.

BERKSHIRE GLASS COMPANY, 1847–1858

16. Raymond E. Barlow and Joan E. Kaiser, *The Glass Industry in Sandwich*, vol. 2 (Atglen, PA: Schiffer Publishing with Barlow-Kaiser Publishing, 1997), 268; this is the most complete history of the development of glass-sand mining in Berkshire County.

17. Ibid.; Berkshire County Registry of Deeds, Adams, Massachusetts, Book 62, 105, 106, 111, 128, 129, 132, 133, 135, 136, 138–43, 159, 180, 212–16, 218, 258, 263, 287; Book 63, 59, 75, 77, 79, 145; Book 64, 99, 157, 277, 279, 309, 337. *Ninth United States Census, 1870* for Pittsfield, Massachusetts (Washington, D.C.: National Archives, Records of the Bureau of the Census, 1870), 166, 180 [hereinafter *Ninth Census, 1870*]; *Tenth United States Census, 1880* for Pittsfield, Massachusetts (Washington, D.C.: National Archives, Records of the Bureau of the Census, 1880), 34, 35 [hereinafter *Tenth Census, 1880*].

18. The initial capital investment in stock was $100,000. *Berkshire County Whig*, "Legislative Acts," February 25, 1847, 3; *Berkshire County Whig*, "Legislature," March 25, 1847, 3; *Emancipator* [Boston], "Massachusetts Legislature," March 31, 1847, 3; Hamilton Child, *The Gazetteer of Berkshire County* (Pittsfield, MA: 1885), 30–31; Harding, "Glass Manufacture," 42.

19. In 1849, a commercial notice announced that shares in the Berkshire Glass Company were selling for the hefty amount of $13; *Salem [Massachusetts] Gazette*, "Monetary Affairs: Sales of Stock in Boston," November 2, 1849, 3. Thomas J. Lobdelle, a Boston broker, became head of the company locally, with William Fuller as the sand agent; Browne, "Over Pathways," 10. In an 1853 directory, the company reported capital of $30,000, with Thomas J. Lobdell as president and William A. Hayes as clerk, treasurer and agent and an office in the Merchants Bank Building; George Adams, *The Massachusetts Register for the Year 1853* (Boston, 1853), 190, 248. Thomas J. Lobdell died in November 1853; *New York Times*, "Death of Thomas J. Lobdell," November 12, 1853, 1. On January 26, 1854, Berkshire

Glass Works stock was trading in Boston at $3.13 and at $3.75 a week later; *Boston Daily Atlas*, "Monetary and Business Affairs," January 26, 1854, 2, and January 30, 1854, 2. Later in the year, in September, the company published notices that it had capital of $80,000; the directors were William Underwood, president, and William S. Lincoln, Nathan Carruth and Peter Harvey; *Pittsfield Sun*, "Berkshire Glass Company [advertisement]," September 7, 1854, 3. The valuation notice of the following year indicated the same amount of capital and officers Harvey, Carruth, Charles L. Thayer and Paul Simpson; *Pittsfield Sun*, "Berkshire Glass Company [advertisement]," September 6, 1855, and September 13, 1855, 3. Stock was trading at $3 in August of that year; *Boston Daily Atlas*, "Monetary and Business Affairs," August 16, 1855, 2. In 1856, Harvey, Carruth, Thayer and Simpson were joined by F. Skinner; *Pittsfield Sun*, "Berkshire Glass Company [advertisement]," November 27, 1856, 3. The precise identities of these men have not been determined, but a later description states that the "stock was principally taken in Boston"; Child, *Gazetteer*, 31. The same group appears in the 1857 announcement; *Pittsfield Sun*, "Berkshire Glass Company [advertisement]," September 10, 1857, 3. Other than these notices, these names do not appear in any other accounts of the factory's history.

20. *Pittsfield Sun*, "[Sand Lake, N.Y., Glass Works]," January 6, 1853, 3. This article noted that the Foxes were insured for $4,000 (about $115,000 today) and that the loss exceeded that by $3,000. This article mistakenly lists the owners as O.R. and S.H. Fox. The Foxes also owned a window glass factory at Durhamville, New York (in Oneida County), purchased in 1845.

21. Child, *Gazetteer*, 31; Harding, "Glass Manufacture," 40, 42.

22. *Pittsfield Sun*, "[A New Post-Office]," November 17, 1853, 2. The New York Post Office boasted that it could deliver mail that was addressed simply to "Glass Works, Berkshire"; "The New York Post Office," *Scribner's Monthly* 16 (May 1878): 59.

23. Richard V. Happel, "Business Only Survival of One-Time Big Glass Works: Sand from County Shipped to Tel Aviv, Desert City," *Berkshire Eagle*, April 30, 1948, 33.

24. "Berkshire Crystal," *Hours at Home: A Popular Monthly of Instruction and Recreation* 11 (October 1870): 524.

25. In 1869, one of the first cylinders of glass made at the factory in 1852 was donated to the Pittsfield Young Men's Association; *Pittsfield Sun*, March 25, 1869, 2. It has not been located.

26. *Pittsfield Sun*, "Plate Glass," December 29, 1853, 2.

27. Josiah Gilbert Holland, *History of Western Massachusetts* (Springfield, MA: Samuel Bowles & Co., 1855), 373. All monetary conversions are according to the Consumer Price Index by the calculator at MeasuringWorth.com, for which 2009 was the most current year at the time of this writing.

28. *Pittsfield Sun*, "45th Annual Fair of the Berkshire Agricultural Society: Premiums Awarded, Agricultural Implements and Mechanical," October 11, 1855, 2; the prize was two dollars (slightly over fifty dollars today).

29. *Pittsfield Sun*, "Botsford's 'We Wag' Time Piece," January 7, 1856, 2.

30. Harding, "Glass Manufacture,"43; Holland, *History*, 373.

31. Harding, "Glass Manufacture," 43.

32. George Brown Tindall and David E. Shi, eds., *America: A Narrative History*, vol. 1 (New York: W.W. Norton and Company, 1999), 707–8.

33. Harding, "Glass Manufacture," 43.

34. *Pittsfield Sun*, "Berkshire Glass Company" and "Valuable Glass Works for Sale," September 10, 1857, 3; repeated on September 17, 3, and September 24, 3.

35. *Pittsfield Sun*, "Berkshire Glass Company," March 18, 1858, 3; Gaffield, *Glass Journal*, vol. 1, 260. The Lenox glassworks, about ten miles south of Berkshire Village, was a large plate-glass factory that operated between 1855, when Richmond leased it, and 1872. Richmond stayed with Lenox for about a year; Weeks, *Report*, 98. Richmond took over the Cheshire Glass Works in November 1862; *Pittsfield Sun*, "The Cheshire Glass Works," November 27, 1862, 3.

36. Gaffield, *Glass Journal*, vol. 1, 18–19, 272; *Pittsfield Sun*, "[Berkshire Glass Works]," July 22, 1858, 2. Gaffield noted that Page & Robbins had purchased the works "at a very low figure, probably at from ⅓ to ¼ of its cost"; Thomas Gaffield, *Notes on Glass*, vol. 4, Thomas Gaffield Papers (MC 139), Institute Archives and Special Collections, MIT Libraries, Cambridge, Massachusetts, 163.

37. *Boston Directory for the Year 1858* (Boston: Adams, Sampson & Co., 1858), 32.

PAGE & ROBBINS, 1858–1863

38. T[homas] G[affield], "A Deserving Glass Manufacturer," *Crockery and Glass Journal* 7 (March 7, 1878): 17.

39. *Second Annual Report of the Government* (Boston: Boston Board of Trade, 1856), 166. In 1859, Page & Robbins relocated to 189 and 191 State, where it stayed for about twenty years; Damrell V. Moore and George Coolidge, *Boston Almanac* (Boston: Brown, Taggard & Chase, 1859), 133; this publication gives a second address at 86 and 88 Central Street. The store moved to 118 Milk Street in the early 1880s. The final address was 451 Atlantic Avenue, starting in 1893; George Gould, *Historical Sketch of the Paint, Oil, Varnish and Allied Trades of Boston since AD 1800* (Boston: Paint and Oil Club, 1914), 137–41.

40. Lorenzo Sabine, *Fourth Annual Report of the Government* (Boston: Boston Board of Trade, 1858), 228; Lorenzo Sabine, *Seventh Annual Report of the Government* (Boston: Boston Board of Trade, 1861), 191.

41. Lorenzo Sabine, *Twelfth Annual Report of the Government* (Boston: Boston Board of Trade, 1866), 137.

42. *Correspondence in regard to the South Boston Flats between the New York and New England Railroad Company, and the Harbor and Land Commissioners of Massachusetts* (Boston: Alfred Mudge & Son, 1880), 110; and *Journal of the House of Representatives of the Commonwealth of Massachusetts* (Boston: Wright & Potter, 1875), 110; *Twenty-third Annual Report of the Board of Trustees of the Free Public Library of the Town of Watertown, Massachusetts* (Watertown: Fred. G. Barker, 1891), 15; Frank B. Goodrich, *The Tribute Book* (New York: Derby & Miller, 1865), 395.

43. Gaffield, *Glass Journal*, vol. 3, 3, 227.

44. Berkshire County Registry of Deeds, Adams, Massachusetts, Book 62, 105, 106, 111, 128, 129, 132, 133, 135, 136, 138–43, 159, 180, 212–16, 218, 258, 263, 287; Book 63, 59, 75, 77, 79, 145; Book 64, 99, 157, 277, 279, 309, 337; Ellen Raynor and Emma Petitclerc, *History of the Town of Cheshire* (Holyoke, MA: Clark W. Bryan & Co., 1885), 150. The Gordon sand beds were to become the largest in the state.

45. Gaffield, *Glass Journal*, vol. 1, 19. In 1859 and 1861, the company's annual financial statement continued to record that the capital was $80,000 (slightly over $2 million today); *Pittsfield Sun*, "Berkshire Glass Company," September 1, 1859, 3, repeated on September 8 and September 15; *Pittsfield Sun*, "Berkshire Glass Company," August 22, 1861, 3, repeated on August 29 and September 5.

46. *Ninth Exhibition of the Massachusetts Charitable Mechanic Association* (Boston: Press of Geo. C. Rand & Avery, 1860), 36.

47. *Pittsfield Sun*, "Obituary [James Raybold]," February 4, 1869, 4.

48. Crystal, "Berkshire," *Pittsfield Sun*, May 27, 1874, 3.

49. Gaffield, *Notes on Glass*, vol. 4, 163.

50. Gaffield, *Glass Journal*, vol. 1, 19.

51. Michael Cable, trans., *Bontemps on Glassmaking: The Guide du Verrier of Georges Bontemps* (Sheffield, UK: Society of Glass Technology, 2008), 102–7.

52. *Eighth United States Census, 1860* for Lanesborough, Massachusetts (Washington, D.C.: National Archives, Records of the Bureau of the Census, 1860) [hereinafter *Eighth Census, 1860*].

53. *Pittsfield Sun*, "Berkshire Glass Company," April 4, 1861, 3.

54. *Berkshire Eagle*, "The Glass Works," April 11, 1861, 2. Harding was described as a partner by Thomas Gaffield, *Glass Journal*, vol. 1, 18. Interestingly, in August the *Pittsfield Sun* ran an advertisement for the sale at auction of personal property and real estate of the Berkshire Glass Company; *Pittsfield Sun*, "Auction Sale," August 29, 1861, 3. Given Page & Robbins' curt assertion that its glass factory was not called Berkshire Glass Company, this sale was probably for whatever land and other goods Page & Robbins had not purchased.

55. *Pittsfield Sun*, January 9, 1862, 3.

56. Frances S. Martin, *Lanesborough, Massachusetts: Story of a Wilderness Settlement, 1765–1965* (Pittsfield, MA: Eagle Printing, 1965), 70.

57. Gaffield, *Glass Journal*, vol. 1, 161; Henry Chance, "On the Manufacture of Crown and Sheet Glass," *Journal of the Society of Arts* (February 15, 1856): 226. "On the Manufacture of Window Glass," *The Pharmacist* 7 (December 1874): 358, stated that English pots were five feet high and five feet in diameter. Page's English pots were slightly smaller: forty-eight inches in diameter at the top, forty-two inches at the bottom and thirty-six inches deep. The American pots were thirty-eight inches in diameter at the top, thirty-four inches at the bottom and thirty inches deep. These are outside dimensions; the walls of the pots were two and a half to three and a half inches thick; Gaffield, *Glass Journal*, vol. 3, 5. "Melting Furnace," in *Knight's New Mechanical Dictionary* (Boston: Houghton, Mifflin, 1884), 593.

Page & Harding, 1863–1883

58. *Pittsfield Sun*, "[Old Glass Works]," August 6, 1863, 2; Gaffield, *Glass Journal*, vol. 1, 18–19, 172.

59. "Rust" on glass was a permanent milky deposit on the surface that occurred when glass had been stored in wooden crates and cushioned with hay. It was caused by acids in the hay reacting with the glass, but this was unknown at the time. The condition fascinated Gaffield and bedeviled Page until about 1865. Gaffield recorded Page's problems with rust frequently; see Gaffield, *Glass Journal*, vol. 1, 20, 172, 193–94, 261, 268, 270, 272–74; vol. 2, 30, 91; vol. 3, 113.

60. Repairs included, for example, the settlement and subsequent rebuilding of the chimney of the glass house in 1864; Gaffield, *Glass Journal*, vol. 1, 269.

61. Robbins was apparently not a "glass man" but a businessman. He moved from Boston to Madison, Wisconsin, to run a woolen mill. By 1900, he was living in Denver, Colorado, where he was retired. Our thanks to Katherine A. Gardner-Westcott, local history librarian at the Watertown (Massachusetts) Free Public Library, for her research on Robbins.

62. Rollin Cooke, *Historic Homes and Institutions and Genealogical and Personal Memoirs of Berkshire County, Massachusetts*, vol. 2 (New York: Lewis Publishing Co., 1906), 356; *Berkshire Eagle*, "Two Prominent Citizens of Pittsfield Die the Same Night," May 20, 1908, 2.

63. *Berkshire Eagle*, "[Mr. William G. Harding]," April 18, 1861, 2.

64. *Berkshire Eagle*, "Dissolution" and "Copartnership," September 9, 1863, 3.

65. *Pittsfield Sun*, July 2, 1873, 2; *Pittsfield Sun*, August 20, 1873, 2; *Crockery and Glass Journal* 5 (March 22, 1877): 16; *American Pottery and Glassware Reporter* (July 1881), in [Collection of papers], 1874–1929, compiled by J. Stanley Brothers, Collection of the Rakow Library, the Corning Museum of Glass [hereinafter Brothers Papers].

66. *Pittsfield Sun*, "The Old Town Halls," April 11, 1867, 2; "Two Prominent Citizens," 2. This building had been located on the site of St. Stephen's Episcopal Church on Park Square; *Proceedings in Commemoration of the Organization in Pittsfield, February 7, 1764, of the First Church of Christ, Pittsfield, Massachusetts* (Pittsfield, MA, 1889), 64.

67. "Two Prominent Citizens," 2. His children were Hope, Malcolm, Isabel, Harriet and George.

68. *Pittsfield Sun*, "Narrow Escape," June 6, 1867, 2; *Pittsfield Sun*, "Lanesborough," June 30, 1870, 2; *Pittsfield Sun*, April 16, 1873, 2.

69. *Pittsfield Sun*, September 10, 1868, 2.

70. Leo Lincoln and Lee Drickamer, *Postal History of Berkshire County, Massachusetts, 1790–1981* (Williamstown, MA, 1982), 49.

71. Class of 1857, *Report of 1872*, 11–12, Williams College Archives and Special Collections, Williamstown, Massachusetts [hereinafter Williams College Archives].

72. "Two Prominent Citizens," 2.

73. *Pittsfield Sun*, "Local Intelligence," January 18, 1866, 2; *Pittsfield Sun*, December 31, 1868, 2; *Pittsfield Sun*, "Pittsfield Branch Bible Society," December 21, 1871,

2. Pontoosuc is a neighborhood of Pittsfield, named for its proximity to a lake of the same name.

74. *Pittsfield Sun*, December 29, 1865, 2.

75. *Pittsfield Sun*, "Local Intelligence," October 15, 1868, 1; *Pittsfield Sun*, "List of Premiums Awarded," October 15, 1868, 2.

76. *Pittsfield Sun*, November 19, 1868, 2; *Pittsfield Sun*, "Local Intelligence," July 15, 1869, 2; *Pittsfield Sun*, "Local Intelligence," July 22, 1869, 2; *Pittsfield Sun*, "The Albany Institute," September 28, 1871, 2; *Proceedings of the Albany Institute*, vol. 1, part 1 (Albany, NY: 1871), 248–56; *Pittsfield Sun*, "An Historical Account of the Albany Institute Visit to the Berkshire Glass Works," October 19, 1871, 2, reprinted in *Glass Club Bulletin* (Winter 2000–1): 8; *Pittsfield Sun*, January 21, 1872, 3. One of his obituaries stated that Harding presented seventeen papers on "wide topics of the day and important social questions" to the Monday Evening Club, 16; Williams College Archives.

77. *Pittsfield Sun*, "Terrible and Probably Fatal Accident to Mrs. Wm. G. Harding," January 7, 1874, 2; "In Memoriam: Nannie Campbell Harding, January 17, 1874," funeral program, Archives of First Congregational Church, Pittsfield, MA; *Pittsfield Sun*, "Death of Mrs. Wm. G. Harding," January 21, 1874, 2.

78. "Two Prominent Citizens," 2; "William Harding Writes," *Class of 1857*, April 20, 1907, Williams College Archives, 18; "William G. Harding," *Class of 1857*, Williams College Archives.

79. *Pittsfield Sun*, "[Page & Harding]," September 7, 1865, 2.

80. *Pittsfield Sun*, "Lanesborough," November 15, 1866, 2.

81. *Pittsfield Sun*, "[Page & Harding's Berkshire Glass Works]," September 14, 1865, 2.

82. Gaffield, *Glass Journal*, vol. 2, 110.

83. Ibid., 121. *Ninth Census, 1870* reports only seventy-nine who were directly employed at the glassworks. Twenty were English (eleven of them were blowers and six gatherers). But this was only the population of Berkshire Village. Others may have lived elsewhere in Lanesborough, Cheshire or Pittsfield.

84. Gaffield, *Glass Journal*, vol. 2, 121; *Pittsfield Sun*, "Lanesborough," November 15, 1866, 2.

85. Bontemps warned that glass factories should be located far from large towns or cities because they were "not favourable to discipline and good order among the workers"; Cable, *Bontemps*, 171.

86. Gaffield, *Glass Journal*, vol. 1, 260–61.

87. Ibid., 260, 269. The first school had been built in 1855, however; Anna Fuller Bennett, *History of the Berkshire Union Chapel* (Pittsfield, MA, 1934), 5.

88. *Pittsfield Sun*, November 13, 1872, 2.

89. Gaffield, *Glass Journal*, vol. 3, 5. There is evidence to suggest that intemperance was exacerbated by the company, which sold alcohol or paid its workers with it from the company store; F.M. Gessner, "American Glass Workers," *The Chautauquan* 13 (June 1891): 322.

90. *Ninth Exhibition*, 36.

91. Raymond McGrath and A.C. Frost, *Glass in Architecture and Decoration* (London: Architectural Press, 1961), 46.

92. Frank S. Brockett, "An Empire Builder: Letter to the Editor," *Berkshire Eagle*, August 26, 1938, 4; James Gillinder, "Glass Manufactures in America," in *The Americana: A Universal Reference Library*, edited by Frederick Converse Beach (New York: Scientific American Compiling Dept., 1912).

93. *Pittsfield Sun*, September 2, 1869, 2.

94. Gaffield, *Glass Journal*, vol. 3, 6.

95. *Memphis Daily Avalanche*, October 27, 1867, 2; *Memphis Daily Avalanche*, January 8, 1868, 2.

96. *Returns from Military Posts, 1800–1916*, National Archives and Records Administration, Washington, D.C.

97. Bliss Perry, "Morris Schaff: A Memoir," *Proceedings of the Massachusetts Historical Society* 64 (Boston: Massachusetts Historical Society, 1932): 3–8. Schaff was also an author and wrote several books on his native Licking County, Ohio.

98. *Official Register of the Officers and Cadets of the U.S. Military Academy* (West Point, NY: United States Military Academy, June 1862), 9.

99. *Pittsfield Sun*, "Up County Items," June 24, 1872, 2. Despite his court-martial, he became brigadier general of the Massachusetts Militia in 1880 (an honorary position), and he received a military pension. He was not dishonorably discharged from the service, and his prison term was never mentioned in any of the obituaries or writings about him. He received several doctoral degrees in late life: an LLD from Williams College in 1913 and a LittD in 1914 from Otterbein University; Perry, "Morris Schaff," 3–8.

100. Happel, "Business Only Survival," 33. Schaff is unnamed but identified as being "married to the owner's daughter."

101. Weeks, *Report*, 35n.

102. Cable, *Bontemps*, 170.

103. *The Glass Industry: Report on the Cost of Production of Glass in the United States* (Washington, D.C.: Department of Commerce, 1917), 12–13.

104. *Pittsfield Sun*, October 19, 1871, 2; *Proceedings of the Albany Institute*, 257; Harding, "Glass Manufacture," 43. The reference is to Birmingham, England, a center of industrial production in the nineteenth century.

105. "Berkshire Crystal," 524. Note: this is a quote in the article. This was probably said by Harding. Some glass furnaces used peat, but Bontemps felt it was inadequate; Cable, *Bontemps*, 133. Peat was useful for other tasks, such as drying wood, calcining lime for use in batch or firing bricks;

106. In 1862, the one eight-pot furnace consumed wood from the surrounding forests, costing $1.75 per cord (presumably in labor only, about $47.00 today). If purchased from others, the wood cost between $2.25 and $2.50 per cord (between $60.00 and $67.00 today) and then still had to be split; Gaffield, *Glass Journal*, vol. 1, 18–19. Within a year, however, the factory had begun to use soft coal (bituminous) for fuel, which cost $5.50 per ton (about $120.00 today); Gaffield, *Glass Journal*, vol. 1, 161; "Berkshire Crystal," 524. Coal had been used for fuel in American furnaces since 1796, when a Pittsburgh furnace adopted it instead of wood, but this was as much because of the proximity of coal deposits to the city as it was a matter of coal being a better fuel; *The Glass Industry*, 11. Bituminous

coal was superior to anthracite (hard coal); Cable, *Bontemps*, 132–33. In 1863, Berkshire began burning hard coal (anthracite) in addition to wood; *Pittsfield Sun*, "[Berkshire Glass Works]," June 25, 1863, 2. They used both coal and wood through the end of the decade; *Pittsfield Sun*, "[Page & Harding's Berkshire Glass Works]," September 14, 1865, 2, reported that the company used $36,000 worth of both coal and wood; *Pittsfield Sun*, "Lanesborough," November 15, 1866, 2, reported that 1,500 cords of wood were consumed every year. One ton of coal cost about $10.00, according to the latter article. For the year 1873, the furnaces consumed sixteen million bushels of coke; *Pittsfield Sun*, November 12, 1873, 2. Two months later, it was reported that the factory used fifteen tons of coal per week, enabling it to ship a train-car load of glass every day; *Pittsfield Sun*, January 7, 1874, 2.

107. Crystal, "Berkshire," *Pittsfield Sun*, May 27, 1874, 3.

108. Gaffield, *Glass Journal*, vol. 3, 44; *Crockery and Glass Journal* 1 (March 20, 1875): 5; *Pittsfield Sun*, August 14, 1875, 4; *Crockery and Glass Journal* 2 (September 9, 1875): 17; *Crockery and Glass Journal* 2 (September 23, 1875): 9. For an explanation of the blast furnace, see "Blast Gas Furnace" in *Knight's New Mechanical Dictionary* (Boston: Houghton, Mifflin, 1884), 105; and Robert Linton, "Continuous Operation in the Manufacture of Window Glass," *Engineering Magazine* 16 (November 1898): 244.

109. "The Berkshire Glass Co.," unidentified publication, September 23, 1875, Gaffield Papers, Scrapbook, vol. 1, Thomas Gaffield Papers (MC 139), Institute Archives and Special Collections, MIT Libraries, Cambridge, Massachusetts, 92.

110. Gaffield, *Glass Journal*, vol. 3, 44.

111. Cable, *Bontemps*, 123–24; *The Glass Industry*, 18.

112. It processed its own gas by burning coal, consuming thirty-five tons in a week; *Pittsfield Sun*, "Cheshire," April 22, 1886, 8; "Gas Generating Furnace," *Knight's New Mechanical*, 384–88; "Glass Furnace," *Knight's New Mechanical Dictionary*, 403–4.

113. "Pot," *Knight's New Mechanical Dictionary* (Boston: Houghton, Mifflin, 1884), 714.

114. The 1860 pots were oval and measured twenty-one inches by twenty-five and a half inches at the bottom and thirty-one and a half inches by thirty-four and a half inches at the top; Gaffield, *Glass Journal*, vol. 1, 19. Gaffield, "Remarks of Messrs. Page & Raybold," August 19, 1861, 5.

115. Gaffield, "Remarks of Messrs. Page & Raybold," August 19, 1861, 5–6; Gaffield says that about sixty pots' worth of glass broke, but half of the glass in these pots was able to be saved.

116. Gaffield, *Glass Journal*, vol. 2, 175; Gaffield, "Remarks of Messrs. Page & Raybold," August 19, 1861, 5.

117. Gaffield, *Glass Journal*, vol. 3, 236–37.

118. Gaffield, Scrapbooks, "Remarks of Messrs. Page & Raybold," June 20, 1861, 1. It is not clear whether this cost was for raw materials alone or included fuel and labor.

119. Gaffield's figures are actually that Stourbridge sold for $80.00 to $90.00 per metric ton (2,240 pounds) and Missouri for $1.60 to $1.70 per hundred pounds.

120. Gaffield, *Glass Journal*, vol. 2, 36.

121. "Berkshire Crystal," 524; *Proceedings of the Albany Institute*, 252. Cheltenham clay had supplanted Stourbridge clay by the late 1850s; Heinrich Ries and Henry Leighton, *History of the Clay-Working Industry in the United States* (New York: John Wiley & Sons, 1909), 120–21.

122. *Pittsfield Sun*, "Local Intelligence," July 22, 1869, 2.

123. C.R. Barns, *The Commonwealth of Missouri: A Centennial Record* (St. Louis, MO: Bryan, Brand & Co., 1877), 604, 607, 611, 665; F.W. Beers, *Topographical Map of Pittsfield, Berkshire Co., Mass.* (Pittsfield, MA: R.T. White & Co., 1876).

124. "Berkshire Crystal," 524; *Pittsfield Sun*, "Local Intelligence," July 22, 1869, 2.

125. Gaffield, *Notes on Glass*, vol. 1, 127; also Thomas Gaffield Papers, Scrapbooks, Bin 2, "Remarks of Messrs. Page & Raybold, June & August, 1861/Remarks of Mr. Hayes on Rust," August 19, 1861, 1. Note: this handwritten notebook contains notes of conversations held on different dates; the page numbering starts over for each new date. James Raybold (1808–1868) was manager of the Berkshire Glass Works from 1858 to 1868; "Obituary," unidentified newspaper clipping, Gaffield papers, Scrapbooks, vol. 1, 34.

126. "Berkshire Crystal," 524; Benjamin Biser, *Elements of Glass and Glass Making* (Pittsburgh: Glass and Pottery Publishing Co., 1899), 74; W.K. Brownlee and A.F. Gorton, "The Manufacture and Treatment of Glass Melting Pots," *Journal of the American Ceramics Society* 4 (February 1921): 97.

127. *Pittsfield Sun*, January 7, 1874, 2.

128. Gaffield noted in 1879 that all of the blowing was done by Belgians, *Glass Journal*, vol. 3, 244. He also noted that they had a different method of blowing the glass than Americans and Englishmen, which he described, 244–45. Cable, *Bontemps*, 169.

129. *Tenth Census, 1880.*

130. *The Census of Massachusetts: 1875. Volume II: Manufactures and Occupations* (Boston: Albert J. Wright, 1877), 12, 474–75.

131. It has been suggested that in most glass factories, the use of scrip was discontinued in the 1880s due to the rise of the unions; Jane Shadel Spillman, "Glasshouse Money—A Real Medium of Exchange (Part Two)," *The Glass Club Bulletin of the National American Glass Club* (Winter 2005): 16. The scrip used by Page & Harding had the name "Berkshire Crystal Glass Works" printed on it. The firm was never incorporated under this name but seems to have it used interchangeably with "Berkshire Glass Works" and "Page, Harding."

132. Happel, "Business Only Survival," 33.

133. *Berkshire Eagle*, "[Letter to the Editor from J.M. Linnehan]," January 19, 1939, 4.

134. Happel, "Business Only Survival," 33; *Berkshire Eagle*, September 17, 1968, 8; Martin, *Lanesborough*, 69.

135. "Glass and Glassware: The Progress of Art in this Industrial Interest, Number Four," *Boston Journal of Commerce* (February 15, 1873), in Gaffield, Scrapbooks, vol. 1, 67.

136. Ibid.

137. Gaffield, *Glass Journal*, vol. 3, 4–6.

138. *The Census of Massachusetts: 1875. Volume II*, 12, 474–75; Gaffield, *Glass Journal*, vol. 2, 261.

139. Gaffield, *Glass Journal*, vol. 3, 5.

140. *Boston Daily Globe*, "New England News by Mail," January 22, 1878, 3.

141. Gaffield, *Glass Journal*, vol. 3, 184, 191.

142. Massachusetts, vol. 5, 201, R.G. Dun & Co. Collection, Baker Library Historical Collections, Harvard Business School; Gaffield, *Glass Journal*, vol. 3, 183–84, 190–91; *Crockery and Glass Journal* 7 (April 25, 1878): 16; "Industrial Notes," *Crockery and Glass Journal* 8 (August 29, 1878): 30.

143. Massachusetts, vol. 5, 201, and vol. 6, 413, 525, 620, 646, R.G. Dun & Co. Collection, Baker Library Historical Collections, Harvard Business School.

144. "Trade Notes," *Crockery and Glass Journal* 9 (February 13, 1879): 24; *North Adams Transcript*, "Berkshire Glass," September 4, 1879, 1.

145. *American Pottery and Glassware Reporter*, October 9, 1879, in Brothers Papers. This is the only reference to there being five furnaces; the remains of only three survive today.

146. *North Adams Transcript*, "Berkshire Glass," September 4, 1879, 1. This was later advertised: "Rendle's Patent System of Glass Roofing [advertisement]," *American Architect and Building News* 10 (May 7, 1881): xiii; Thomas Gaffield, "The Uses of Glass in Photography," *The Photographic Times and American Photographer* 11 (August 1881): 313; *History of Berkshire County, Massachusetts with Biographical Sketches of Its Prominent Men*, 626. They were still making it as of 1891; *Pottery and Glassware Reporter*, September 10, 1891, in Brothers Papers.

147. *Crockery and Glass Journal* 11 (May 4, 1880): 10. This calculation of earnings is based on the *Massachusetts Census of 1885*, 1053, which reported that a box of glass averaged $2.50; and the *Massachusetts Census of 1875*, 474, which stated that the factory operated three hundred days a year.

148. *North Adams Transcript*, "Berkshire Glass," September 4, 1879, 1.

149. *Crockery and Glass Journal* 11 (May 4, 1880): 10; Amos D. Snow, "Berkshire Sunday School," report to First Church, Pittsfield, Mass., July 12, 1886.

Berkshire Glass Company, 1883–1899

150. Massachusetts, vol. 6, 525, R.G. Dun & Co. Collection, Baker Library Historical Collections, Harvard Business School. It was capitalized for $125,000 ($2.77 million today) with stock issued at a par value of $100, but all shares were owned by Harrison Page, William G. Harding and their families.

151. *History of Berkshire County*, 626.

152. *Pittsfield Sun*, "Lenox Furnace," November 8, 1886, 8.

153. *Pittsfield Sun*, "Plate Glass," December 29, 1853, 2; *Boston Directory for the Year 1858* (Boston: Adams, Sampson & Co., 1858), 32.

154. Child, *Gazetteer*, 32; *Pittsfield Sun*, "The Lenox Plate Glass Co. vs. Wm. E. Dodge, for $600,000.—Charges of Fraud Made," May 29, 1878, 1.

155. *Pittsfield Sun*, "Lenox Furnace," March 10, 1887, 8; Office of the Counsel to the Corporation, *Report for Year Ending December 31, 1890* (New York: City of New York Law Department, 1890), 246–47.

156. *North Adams Transcript*, "Berkshire Glass," September 4, 1879, 1; Francis B.C. Bradlee, *The Boston and Lowell Railroad, The Nashua and Lowell Railroad, and the Salem and Lowell Railroad* (Salem, MA: Essex Institute, 1918), 44.

157. *Seventh United States Census, 1850* for Berkshire Village, Lanesborough, Massachusetts (Washington, D.C.: National Archives, Records of the Bureau of the Census, 1850).

158. Anna Fuller Bennett, *History of the Berkshire Union Chapel* (Pittsfield, MA: 1934), 8.

159. *Massachusetts Ploughman*, "Among the Berkshire Farmers," August 13, 1870, 17.

160. *New York Times*, "No Protection in This: The High-Tariff Fraud in Window Glass Making," October 8, 1888, 1.

161. "Cheshire Sand Works and Ore Beds," *Crockery and Glass Journal* (September 10, 1875): 20.

162. "No Protection in This," 1. The report on the glass industry in the 1890 U.S. Census reported no children employed in Massachusetts glass factories; see Joseph D. Weeks, "Glass," in Department of the Interior, Census Office, *Report on the Manufacturing Industries in the United States at the Eleventh Census: 1890*, Part III, Selected Industries (Washington, D.C.: United States Census Bureau, 1895): 328.

163. *Eighth Census, 1860.*

164. Bennett, *Berkshire Union Chapel*, 5, 6.

165. Ibid., 6.

166. Ibid.

167. *Ninth Census, 1870*; Crystal, "Berkshire," *Pittsfield Sun*, May 27, 1874, 3.

168. Bennett, *Berkshire Union Chapel*, 6.

169. *Tenth Census, 1880.*

170. Snow, "Berkshire Sunday School," 8.

171. Bennett, *Berkshire Union Chapel*, 8–12; *Pittsfield Sun*, "Berkshire," May 31, 1888, 8. The building still stands at the intersection of State Route 8 and Old State Road in Lanesborough.

172. *Pittsfield Sun*, "Berkshire: News from Crystal City," May 31, 1888, 2.

173. *Pittsfield Sun*, September 23, 1858, 2.

174. *Pittsfield Sun*, May 29, 1872, 2

175. *Pittsfield Sun*, June 19, 1872, 2.

176. Public Document 34, *Twenty-first Annual Report of the State Board of Health of Massachusetts* (Boston: Wright & Potter, 1890): xii–xiii.

177. *Twelfth United States Census, 1900* for Berkshire Village, Lanesborough, Massachusetts (Washington, D.C.: National Archives, Records of the Bureau of the Census, 1900). There is no U.S. census data available for 1890.

178. "Glass Notes," *Crockery and Glass Journal* (August 18, 1898): 8; *North Adams Transcript*, "Cheshire," January 30, 1899, 8; *North Adams Transcript*, "Cheshire," February 10, 1899, 5.

179. See, for instance, *Boston Globe*, "Trade and Labor," July 17, 1882, 4; and *New York Times*, "Window Glass Workers' Convention," July 17, 1882, 5. The *Tenth*

Census, 1880 reported that by 1880, about nine-tenths of the window-glass workers in the country belonged to the Window-Glass Workers' Association, which was part of the Knights of Labor; William C. Birdsall, "The Problem of Structure in the Knights of Labor," *Industrial and Labor Relations Review* 6 (July 1953): 541. See also Ken Fones-Wolf, *Glass Towns: Industry, Labor, and Political Economy in Appalachia, 1890–1930s* (Urbana and Chicago: University of Illinois Press, 2007), 20–22.

180. See, for instance, *New York Times*, "Foreign Contract Labor," March 14, 1883, 4; and *Boston Globe*, "Bona Fide Protection," March 16, 1883, 2.

181. Gessner, "American Glass Workers," 320.

182. Gillinder, "Glass Manufactures in America."

183. Linton, "Continuous Operation," 244.

184. Gillinder, "Glass Manufactures in America."

185. Public Document 34, *Twenty-First Annual Report of the State Board of Health, Massachusetts* (Boston: Wright & Potter, 1890), xii.

186. Committee on Finance, United States Senate, "Reply of the Berkshire Glass Company, Mass.," *Replies to Tariff Inquiries, Schedule B, Continued, Earths, Earthenware, and Glassware, Bulletin No. 1, Opinions of Collectors of Custom Concerning Ad Valorem and Specific Rates of Duty on Imports* (Washington, D.C., 1894), 169. Window glass imported from Belgium was often less expensive than domestically produced window glass; see, for example, Gaffield, *Glass Journal*, vol. 3, 243.

187. Committee on Finance, United States Senate, "Reply of the Berkshire Glass Company, Mass.," *Replies to Tariff Inquiries, Schedule B, Continued, Earths, Earthenware, and Glassware*, 169; Harding, "Glass Manufacture,"43.

188. *Boston Daily Globe*, "The House Session," February 24, 1876, 2.

189. Harding, "Glass Manufacture," 44.

190. Gessner, "American Glass Workers," 319. This article reports that the number of window glass factories was 88 in 1880 and 158 in 1890, and the number of cathedral-glass factories was 2 in 1880 and 16 in 1890.

191. Weeks, "Glass," 315.

192. *The Industries of Saint Louis* (St. Louis, MO: J.M. Elstner and Co., 1887), 98.

193. *Annual Report of the Secretary of Internal Affairs of the Commonwealth of Pennsylvania, part III, Industrial Statistics*, 28 (1900), 13.

194. Fones-Wolf, *Glass Towns*, 23–24.

195. *Boston Daily Globe*, "Business Troubles," February 13, 1894, 7; *China, Glass & Lamps*, August 15, 1894, in Brothers Papers.

196. Perry, "Morris Schaff," 3–8. He served on the state Gas and Electric Commission for twenty-six years following his time at Berkshire Glass Works; Samuel Schaff, "Morris Schaff: Author/ Soldier/ Historian/ Public Servant," *Historical Times, Newsletter of the Granville, Ohio, Historical Society* (Fall 1990): 1–3.

197. *North Adams Transcript*, "Cheshire," November 12, 1895, 3; *North Adams Transcript*, "Berkshire Glass Works," November 4, 1897, 4.

198. *North Adams Transcript*, "Berkshire Glass Works," November 4, 1897, 4. *New York Times*, "Glassworkers Troubles End," January 9, 1898, 1.

199. "Glass Notes," *Crockery and Glass Journal* (August 18, 1898): 23.

200. *North Adams Transcript*, "Local Intelligence," March 16, 1899, 5; *North Adams Transcript*, "Cheshire," March 17, 1899, 8; *Crockery and Glass Journal* (March 23, 1899), in Brothers Papers.

BERKSHIRE CO-OPERATIVE GLASS COMPANY, 1899–1903

201. There were two Conrad Hineses employed by the glassworks. The first came to work for the factory in the 1860s, first appearing in the *Ninth Census, 1870* when he was twenty-one. He was born in Darmstadt, Germany, in January 1849 and emigrated as a child of six. In 1870, he married Marie Harriet Jackley, born in Sand Lake, New York, in 1852, whose family probably came to Berkshire Village with Albert Fox when he became manager of the factory in 1858 after the Sand Lake Glass Factory burned. Conrad died between 1910 and 1920. His son Conrad was born around 1877. Both worked as flatteners, and either could have been the president of the cooperative. One of them (probably the son) gave a collection of glass samples and whimsies to the Berkshire Museum, Pittsfield, Massachusetts, which is now divided between the museum and the Berkshire Historical Society, Pittsfield.

202. *North Adams Transcript*, "Starting the Glass Works," October 24, 1899, 8.

203. *Washington Post*, "The World of Labor," November 5, 1899, 22.

204. *North Adams Transcript*, "A Co-Operative Glass Works," October 6, 1899, 5.

205. *North Adams Transcript*, "Glass Plant in Operation," October 27, 1899, 8.

206. *New York Times*, "Glass Factories to Close," March 31, 1901, 1.

207. R.M. Smythe, *Obsolete American Securities and Corporations* (New York: 1911), 143.

208. "The Best Material of Its Kind in the World," *The Berkshire Hills: A Historical Monthly* (November 1, 1904): 18.

209. Happel, "Business Only Survival," 33; *North Adams Transcript*, "Trouble in Holy Land May Halt Annual Sand Order," May 12, 1948, 7.

210. *North Adams Transcript*, "Out-of-Towner Buys Old Cheshire Firm," January 5, 1951, 6; *Berkshire Eagle*, "Glass Plays no Part Today in Glass Sand Co. Product," July 6, 1957, 8.

COLORED GLASS

211. *Tenth Exhibition of the Massachusetts Charitable Mechanic Association* (Boston: Wright & Potter, 1865), 106; Edward Dewson, "American Stained Glass," in *Catalogue of the Art Department of the New England Manufacturers' and Mechanics' Institute* (Boston: Cupples Upham & Co., 1883), says that Page began making colored glass in 1869–70.

212. "Beauty in Fire Screens," *Carpentry and Building* 7 (December 1884): 227.

213. Unidentified clipping, 1871, Gaffield Papers, Scrapbooks, vol. 3, 35.

214. The congregation of St. Luke's occupies two different buildings; the Stone Church is one of them. The decorative panels in the middle of the window were made by the Belcher Mosaic Company of New Jersey.

215. *Tenth Exhibition*, 106.

216. "Trade Notes," *Crockery and Glass Journal* 4 (November 2, 1876): 12.

217. Tony Benyon, "The development of Antique and other glasses used in 19th- and 20th-century stained glass," *Journal of Stained Glass* 29 (2005): 190, 196n9; Stanley A. Shepherd, *The Stained Glass of A.W.N. Pugin* (Reading, UK: Spire Books, 2009), 51–58.

218. Benyon, "Development," 190, 196n9; "On the Manufacture of Window Glass," *The Pharmacist* 7 (December 1874): 359; in 1862, it was being made by Wisthoff & Co. in Germany; production began in France in 1864.

219. *North Adams Transcript*, "Berkshire Glass," September 4, 1879, 4.

220. "Glass," *Catalogue of the Thirteenth Exhibition of the Massachusetts Charitable Mechanic Association* (Boston, 1878), 77. The window no longer survives.

221. *North Adams Transcript*, "Berkshire Glass," September 4, 1879, 1.

222. Chance Brothers, *Patterns of Enamelled, Double-Etched, and Stained Enamelled Glass* (Birmingham, England, 1863); Carter Brothers Catalog owned by Julie L. Sloan; Pittsburgh Plate Glass, *Glass: History, Manufacture and its Universal Application* (Pittsburgh: Pittsburgh Plate Glass Company, 1923), 131. This is not to suggest that the designs originated with Page & Harding.

223. The date and origin of the hanging lamp at Chesterwood are unknown, but the enameled glass is water-white, like Berkshire glass. Chesterwood itself was not built until 1901, but the lamp is thought to have been brought to the house from one of French's earlier residences. Our thanks to Anne L. Cathcart, curatorial assistant at Chesterwood.

224. Gaffield, *Glass Journal*, vol. 3, 177.

225. Ibid., 228, 234–35. Schaff's problems with green may also have been due to using copper oxide as a colorant, which was adversely affected by coal fires and by the cullet used in the batch; Cable, *Bontemps*, 275–76. Gaffield had recorded that Page used poplar sawdust in 1875; *Glass Journal*, vol. 3, 114. Bontemps recommended using poplar or alder sawdust; 79, 269–70.

226. Augustus J. Pleasanton, *The Influence of the Blue Ray of the Sunlight and of the Blue Color of the Sky in Developing Animal and Vegetable Life* (Philadelphia: Claxton, Remsen & Haffelfinger, 1877); Pleasanton had published a short pamphlet, *On the Influence of the Blue Color of the Sky in Developing Animal and Vegetable Life*, in 1871, but it was the publication of his book that caused the craze for blue glass.

227. Unidentified clipping, probably from the *Boston Journal of Commerce*, February 6, 1877, Gaffield, Scrapbooks, vol. 2, 42.

228. *New York Tribune*, "General Notes," February 15, 1877, 5.

229. *Crockery and Glass Journal* 5 (March 22, 1877): 15–16; unidentified clipping, Gaffield, Scrapbooks, vol. 2, 42.

230. *Crockery and Glass Journal* 5 (March 22, 1877): 15–16.

231. Ibid.

232. *New York Tribune*, "General Notes," February 15, 1877, 5.

233. Gaffield, *Glass Journal*, vol. 3, 165.

234. Unidentified clippings, Gaffield, Scrapbooks, vol. 2, 42–43.

235. Gaffield, *Glass Journal*, vol. 3, 165. Smalt is a frit of soft cobalt glass; Cable, *Bontemps*, 74; Woldemar A. Weyl, *Coloured Glasses* (London: Dawson's of Pall Mall, 1959), 190.

236. Gaffield, *Glass Journal*, vol. 3, 208–9.

237. *American Pottery and Glassware Reporter*, January 31, 1884, in Brothers Papers.

238. *Crockery and Glass Journal* 11 (April 29, 1880): 14; *Crockery and Glass Journal* 11 (May 4, 1880): 10.

239. *Crockery and Glass Journal* 19 (January 10, 1884): 30.

240. *The Industries of Saint Louis*, 98.

241. James Cox, ed., *Missouri at the World's Fair* ([St. Louis, Missouri]: World's Fair Commission of Missouri, 1893), 125.

STAINED-GLASS WINDOWS MADE WITH BERKSHIRE GLASS

242. For MacDonald's life and work, see Lance Kasparian, "The stained glass work of Donald MacDonald of Boston: a preliminary study," *Journal of Stained Glass* 28 (2004): 12–30.

243. Ibid., 16–18.

244. McPherson also provided a figural window of the Good Shepherd in the Parish Hall and a three-lancet window depicting St. Michael the Archangel in the north transept. Removed in 1901, it is now located in the Church of the Immaculate Conception in North Easton.

245. *Crockery and Glass Journal* 11 (May 4, 1880): 10.

246. Paul Norton, *Rhode Island Stained Glass: An Historical Guide* (Dublin, NH: William L. Bauhan, 2001), 212.

247. Contract between Cook, Redding & Co. and Waterloo Library, holograph, February 9, 1882, Collection of the Waterloo Library and Historical Society, Waterloo, New York

248. *Crockery and Glass Journal* 11 (April 29, 1880): 14.

249. This listing is in an undated sales catalogue from W.J. McPherson & Co. that is in the possession of Lance J. Kasparian, Salem, Massachusetts.

250. Henry C. Van Brunt, "Studies in Interior Decoration. XIII," *American Architect and Building News* 2 (July 21, 1877): 232.

251. Samuel West to Trinity Episcopal Church, November 27, 1876, Archives, Trinity Episcopal Church in the City of Boston, Boston, Massachusetts; H. Barbara Weinberg, *The Decorative Work of John La Farge* (New York: Garland Publishing Inc., 1977), 107. See also Julie L. Sloan and James L. Yarnall, "The Art of an Opaline Mind: The Stained Glass of John La Farge," *American Art Journal* 24 (1992): 4–43.

252. We are using the term "Tiffany Studios" here as a catch-all name for all of Louis Comfort Tiffany's various companies that produced stained glass.

253. "Connecticut: Pomfret—Christ Memorial Church," *The Churchman* 7 (May 5, 1883): 485.

254. The church, designed by architect Howard Hoppin (1854–1940), was built as a memorial to Reverend Alexander H. Vinton (d. 1881). The windows are dedicated to members of the Thompson, Clark and Vinton families, who were related by marriage.

255. Retta Lou Weber and Gayle Weber Strange, *Lively Stones: A History of the People Who Built First Presbyterian Church, Galveston, Texas* (Franklin, TN: Providence House, 1993), 51.

256. Martin, *Lanesborough*, 68. For more information on Crowninshield's work, see Gertrude deG. Wilmers and Julie L. Sloan, *Frederic Crowninshield: A Renaissance Man in the Gilded Age* (Amherst: University of Massachusetts Press, 2010).

257. "Designs for Stained-Glass Windows, by Mr. Frederic Crowninshield, Artist, Boston, Mass," *American Architect and Building News* 291 (July 23, 1881): 39; *The Commemoration by the First Church in Boston of the Completion of Two Hundred and Fifty Years since Its Foundation.* (Boston: Hall and Whiting, 1881), 208–10.

258. *Proceedings in Commemoration of the Organization in Pittsfield, February 7, 1764, of the First Church of Christ, Pittsfield, Massachusetts* (Pittsfield, MA, 1889), 38.

259. Although Prentice Treadwell appears to have been well known and respected in his day, there is very little information about him. He opened a studio in Boston in 1883, when it was said that his work was "extremely popular"; *Boston Daily Journal*, "Art in Glass," October 13, 1883, 5. He decorated the Metropolitan Opera House, the "old" Lyceum Theater and Rector's Theater in New York; homes for Senator Stephen B. Elkins in Washington, D.C., and George J. Gould in Lakewood, New Jersey; and several insurance company buildings; "In Bas Relief," *Building Trades Association Bulletin* 4 (May 1903): 84.

260. *First Church of Christ, Pittsfield*, 38.

261. "VIII. The Alumni Memorial Window," *Thirty-Second Annual Report of the Superintendent of Public Instruction of the State of New York* (Albany, NY: Weed, Parsons and Company, 1886), 90–91,199–200; *An Historical Sketch of the State Normal College at Albany, New York and a History of its Graduates for Fifty Years* (Albany, NY: Brandow Printing Co., 1894), 20. The building burned on January 8, 1906: "A feature of the loss is an immense and beautiful stained glass memorial window, given to the college by students and alumni. It was regarded as one of the finest specimens of American colored glass work"; "New York State: State Normal College Burned," *American Education* 9 (February 1906): 360. The window cost $5,000.

262. Isaac Edwards Clarke, "A Notable Art Project," *Art and Industry: Education in the Industrial and Fine Arts in the United States*, part II (Washington, D.C.: U.S. Department of the Interior, Bureau of Education, 1892), 11–12.

263. Louis Comfort Tiffany, "American Art Supreme in Colored Glass," *Forum* 15 (July 1893): 625.

264. *Berkshire Evening Eagle*, "Catholics of Berkshire Village with Pick, Saw and Shovel Built Chapel," December 4, 1935, 8; *Berkshire Eagle*, "Father McMahon Sees Dream of His Boyhood Come True," December 9, 1935, 8; *Berkshire Evening Eagle*, "Grass Now Covers Land Where Once Best Glass in America was Made," December 11, 1935, 8.

Appendix I

265. *Journal of the Honorable Senate of the State of New Hampshire, June Session, 1868* (Manchester, NH: John B. Clarke, 1868), 255.

266. *North Adams Transcript*, "Berkshire Glass," September 4, 1879, 1.

267. Kathryn E. Holliday, *Leopold Eidlitz: Architecture and Idealism in the Gilded Age* (New York: W.W. Norton, 2008), 137–38. One of the authors worked on the restoration and had examples of the original to compare with excavated samples.

268. Martin, *Lanesborough*, 37.

269. *New York Times*, "Catholic Church Matters: The New Building for the Congregation of the Catholic Church of the Sacred Heart," March 30, 1884, 3; *New York Times*, "A New Catholic Church," April 17, 1884, 8. Robert Koch, "The Stained Glass Decades: A Study of Louis Comfort Tiffany (1848–1933) and the Art Nouveau in America," (PhD diss., Columbia University, 1957), 59.

270. Weber and Strange, *Lively Stones*, 51.

INDEX

About the Authors

William Patriquin has had a lifelong admiration for and obsession with glass. Born and raised in Berkshire Village, Bill has been examining the site and compiling research for many years. Prior to starting his career in glass, Bill served in the U.S. Navy as a hospital corpsman, BioMed technician and navy diver. He retired in 1997 as a chief petty officer. Upon returning to the Berkshires, he became a professional stained-glass craftsman and restorer and has worked on windows by Tiffany, La Farge and many others.

Julie L. Sloan is a stained-glass consultant in North Adams, Massachusetts. She has worked in stained glass since 1982 and is the author of *Conservation of Stained Glass in America*. Her BA in art history is from New York University, and her MS in historic preservation is from Columbia University. Her other books include *Light Screens: The Complete Leaded-Glass Windows of Frank Lloyd Wright* (Rizzoli International, 2001) and *Frederic Crowninshield: A Renaissance Man in the Gilded Age* (University of Massachusetts Press, 2010) with Gertrude deG. Wilmers. She has contributed to *A New and Native Beauty: The Art and Craft of Greene & Greene* (Edward Bosley and Anne Mallek, eds., Merrell, 2008) and *Frank Lloyd Wright: Art Glass of the Martin House* (Eric Jackson-Forsberg, ed., 2009). Her stained-glass restoration projects include Saint Thomas Episcopal Church, New York; H.H. Richardson's Trinity Church in Boston; Harvard University's Memorial Hall; Princeton University's Chapel; and the State Houses of Connecticut, Massachusetts, New Jersey and Pennsylvania.

Visit us at
www.historypress.net